USING COMPUTER TOOLS FOR ELECTRIC CIRCUITS

FIFTH EDITION

JAMES W. NILSSON
PROFESSOR EMERITUS
IOWA STATE UNIVERSITY

SUSAN A. RIEDEL
MARQUETTE UNIVERSITY

Using Computer Tools for Electric Circuits

FIFTH EDITION

James W. Nilsson
PROFESSOR EMERITUS
IOWA STATE UNIVERSITY

Susan A. Riedel
MARQUETTE UNIVERSITY

 ADDISON-WESLEY PUBLISHING COMPANY
Reading, Massachusetts • Menlo Park, California
New York • Don Mills, Ontario • Wokingham, England
Amsterdam • Bonn • Sydney • Singapore
Tokyo • Madrid • San Juan • Milan • Paris

Matlab® is a registered trademark of The MathWorks, Inc. To order or receive further information, please contact:

The MathWorks, Inc.
24 Prime Park Way
Natick, MA 01760-1500
Phone: (508) 653-1415
Fax: (508) 653-2997
Email: info@mathworks.com

MicroSim™ Probe® and MicroSim™ PSpice® are registered trademarks of MicroSim Corporation in the United States and other countries throughout the world. All other MicroSim product names, including MicroSim™, are trademarks of MicroSim Corporation. Representations of MicroSim PSpice and other MicroSim products are used by permission of and are proprietary to MicroSim Corporation. Copyright 1995 MicroSim Corporation. To obtain a free examination copy of MicroSim products call 1-800-245-3044 or download from their BBS at 714-830-1550.
Quattro® is a registered trademark of Novell, Inc.
Maple® is a registered trademark of Waterloo Maple Software.

ISBN 0-201-84707-8
8 9 10—DOC—9998

PREFACE

Using Computer Tools for Electric Circuits provides guidelines for employing a variety of computer tools as part of an introductory course in electric circuit analysis using the textbook *Electric Circuits, Fifth Edition*. This supplement looks at five categories of computer tools:

* Circuit simulators;

* Circuit schematic capture and analysis packages;

* Matrix equation solvers;

* Computer spreadsheets; and

* Symbolic equation solvers.

We focus on the application of these tools to the introductory study of electric circuits, and therefore do not exploit the full power of any of the tools discussed. Further, we assume that you are already familiar with using the tools you are interested in — perhaps you learned to use a spreadsheet package in an introductory engineering course, or you used a symbolic equation solver to support your study of calculus. This supplement will not teach you to use any of the tools described. Instead, it will examine how you can use a familiar tool to support your study of electric circuits. If you are not familiar with a particular tool, we encourage you learn about it before using this supplement. We include references for each of the tools in the appropriate chapter.

How can computer tools help you in an electric circuits course? We think there are many advantages to using one or more computer tools to assist the learning process:

* Computer-generated solutions of circuit equations can validate your intuition about a solution and its hand calculation.

* Computer-generated solutions of circuit equations can replace hand calculations when the circuit is complicated and there are many equations. You can thus explore circuits beyond those relatively simple examples suitable for an introductory textbook. Remember that you will still need independent validation that the computer solution makes sense and is correct.

* Once the computer solution is available, each of the tools can provide a visual representation of the solution, which would be tedious to produce without the computer. Visualizing a solution often leads to a deeper and more complete understanding of the circuit behavior which the solution represents.

⁕ Once a computer solution has been validated for one set of circuit parameters, the computer can generate solutions for other sets of circuit parameters. You can then see the effect of changing one or more circuit parameters on the behavior of the circuit without resorting to repeated hand calculations.

Computer tools cannot replace the more traditional means of studying electric circuits. Many of the tools we describe were not specifically designed for electric circuit analysis, and therefore rely on the user to specify the equations that describe the circuit's behavior. All of the tools require the solutions they provide to be independently validated. These tools are meant to assist in the learning process, to deepen your understanding by providing visualization of circuit behavior, and to lighten the computational burden of exploring more complicated circuits.

The disk that accompanies this manual provides several examples of circuit problems whose analysis is assisted by the computer. For more information on how to use the disk, see the file named `readme.doc` in the main directory of the disk. Since this supplement supports *Electric Circuits, Fifth Edition*, most of the examples are drawn from example problems and chapter problems in this parent text. These examples are intended to give insight into the types of circuit problems amenable to computer analysis, but by no means exhaust the possibilities of the computer tools. The remainder of this supplement provides a brief description of each software tool category, a brief description of the particular software package used to illustrate the category, a discussion of the examples on the disk, and a list of suggested circuit problems for you to explore using the computer.

INTEGRATING COMPUTER ANALYSIS INTO INTRODUCTORY CIRCUITS COURSES

We feel that students using *Electric Circuits, Fifth Edition* will benefit from the use of one or more computer tools into the electric circuits course. We use the following methods to support this integration: (1) the textbook example problems or chapter problems adapted for use as examples in this supplement are explicitly identified by number, providing a direct reference to the text; (2) margin notes in the text indicate some topics that you can explore using computer tools; and (3) nearly 150 problems from the text (listed in the *Instructor's Manual* for your easy reference) have been included in this supplement for you to solve using computer tools.

LIST OF EXAMPLES

CIRCUIT SIMULATION

While the other categories of computer tools covered in this supplement have a wide variety of applications, and can be adapted to support the study of electric circuits, the circuit simulation tools described in this chapter and the schematic capture tools described in Chapter 2 are specifically designed to support the study of circuits, both analog and digital. In general, circuit simulators require you to learn a new language, which is used to describe the circuit, the type of analysis to be performed, and the form of the output. Since it is unlikely that you would have learned such a language before embarking on the study of electric circuits, we spend some time explaining the language of a particular circuit simulator, as well as demonstrating some examples of circuit simulation.

A circuit simulator can support your study of electric circuits in many ways. It can verify hand calculation of circuit variables. It can permit you to vary circuit component values and reanalyze the circuit without recalculating by hand. It can support the construction of circuit models for basic devices like op amps and transformers. It can analyze circuits that are too complicated for hand analysis. Many circuit simulators support a graphical interface to the output of the simulation, which permits you to visualize the behavior of the circuit.

The specific circuit simulator used in the examples which follow is MicroSim™ PSpice®, by MicroSim Corporation.[1] MicroSim PSpice

[1]The examples that follow were generated using *The Student Edition of PSpice, Version 5.0.*

is a popular and widely used circuit simulator for many reasons. It has a user-friendly interface, supports graphical exploration of the simulation output using a program called MicroSim Probe®, supports simulation of analog, digital, and mixed circuits, and serves as the simulation platform for MicroSim™ Schematics, which we discuss in Chapter 2. The student version of MicroSim PSpice, used to generate the examples here, can be obtained at no cost by students. We shall limit our applications of MicroSim PSpice to the types of circuit problems discussed in the text. Although MicroSim PSpice is a general-purpose program designed for a wide range of circuit simulation — including the simulation of nonlinear circuits, transmission lines, noise, and distortion — here we discuss only the use of MicroSim PSpice in dc analysis, transient analysis, and steady-state sinusoidal (ac) analysis.

INTRODUCTION TO MICROSIM PSPICE

The general procedure for using MicroSim PSpice consists of three basic steps. In the first step we create a source file for the circuit to be simulated or analyzed. In the second step we enter the source file into the computer, which then runs MicroSim PSpice to simulate the circuit described by the source file and creates an output file. The third and final step is to print or plot the results of the simulation from the output file.

Before we discuss the creation of the source file, some general comments about the MicroSim PSpice format are in order.[2]

1. Each statement in the source file may have several constituent parts. These parts, called fields, must be in a specific order but may appear anywhere in the line. You can separate fields with one or more blanks, commas, or tabs. In certain instances, you can use equal signs or left and right parentheses as separators.

2. A statement in the source file cannot be longer than the 80 characters that can fit within a single line on the computer display. You can continue a longer statement on the next line by including a plus sign (+) in column 1 of that next line, followed by the remainder of the statement.

3. MicroSim PSpice does not distinguish between upper- and lowercase characters. For example, "VIN," "Vin," and "vin" are equivalent. It is the convention in this manual to use uppercase for MicroSim PSpice keywords and lowercase for all other words.

---

[2]*MicroSim PSpice Circuit Analysis, Version 5.0.* MicroSim Corporation, Irvine, CA, 1991.

4. A name field must begin with a letter, "a" through "z" (or "A" through "Z"), but the characters that follow can be letters, numbers, or any of the following: "$," "_," "*," "/," or "%." Names can be up to 131 characters long, but it is good practice to choose a more reasonable length, say eight characters.

5. A number field may be an integer (4, 12, −8) or a real number (2.5, 3.14159, −1.414). Integers and real numbers may be followed by either an integer exponent (7E−6, 2.136e3) or a symbolic scale factor (7U, 2.136k). The symbolic scale factors and the corresponding exponential forms are summarized in Table 1. Letters immediately following a number that are not scale factors are ignored, and so are letters immediately following a scale factor. For example, 10, 10V, 10Hz, and 10A all represent the same number. The same can be said for 2.5M, 2.5MA, 2.5Msec, and 2.5MOhms.

6. The first statement in the circuit description file is the title line, which may contain any type of text. The title line is ignored by MicroSim PSpice, except to label your output once analysis is complete.

7. The last statement in your circuit description file must be .END. Notice that the period is part of the statement.

8. Comment lines are marked by an "*" in the first column. Comments may contain any text and are ignored by MicroSim PSpice.

9. Except for the title line, subcircuit definitions, the .OPTIONS statement with the NOECHO parameter and the .END statement, the order of the statements in the source file does not matter.

TABLE 1.1

MICROSIM PSPICE SCALE FACTORS

SYMBOL	EXPONENTIAL FORM	VALUE
F	1E-15	10^{-15}
P	1E-12	10^{-12}
N	1E-9	10^{-9}
U	1E-6	10^{-6}
M	1E-3	10^{-3}
K	1E3	10^{3}
MEG	1E6	10^{6}
G	1E9	10^{9}
T	1E12	10^{12}

In creating a source file that directs MicroSim PSpice to analyze a circuit, we must do three things. First, we must describe the circuit to be analyzed; second, we must state the type of analysis to be performed; and third, we must specify the desired output. It is convenient therefore to divide the source file into three major subdivisions.

The first subdivision consists of data statements that describe the circuit being simulated. The second subdivision is made up of control statements that describe the type of analysis to be performed. The third subdivision contains the output specification statements, which control what outputs are to be printed or plotted.

Remember that in addition to these major statement subdivisions, the source file must include two other statements. The *first* line of the source file must be a *title statement* and the *last* line of the source file must be an *end statement*.

You can create the source file with any text editor available on your computer, so long as this editor does not insert any control characters. Note that most word processors make extensive use of control characters and so are generally not suited for creating MicroSim

PSpice source files. If your word processor has a configuration in which text can be created without the usual control characters, then you can create MicroSim PSpice source files using this configuration. Also, if you are using MicroSim PSpice within the Control Shell™, the Files menu contains a text editor that you can use to create the source file. Working within the Control Shell has other significant advantages, such as the on-line help manual, which can assist in creating and modifying source files.

We now take a closer look at the three major subdivisions within a MicroSim PSpice source file: the data statements, which describe the circuit, the control statements, which specify the type of analysis, and the output statements, which describe the type and format of the output.

DATA STATEMENTS

MicroSim PSpice is based on nodal analysis. Therefore the first step in describing a circuit to MicroSim PSpice is to number *all* the nodes in the circuit. The reference node *must* be numbered zero (0). The remaining nodes *must be numbered with non-negative integers* but they need not be sequential. Once you have numbered all the nodes in a circuit, you can completely describe the circuit by identifying the type of element that is connected between the nodes. In addition to describing the type of element, we must also specify its numerical characteristics. MicroSim PSpice places several restrictions on the topological characteristics of a circuit. Therefore, in describing a circuit we must make sure that (1) every node has at least two connections; (2) each node in the circuit has a dc path to the reference node; (3) the circuit does not contain loops of either voltage sources or inductors; and (4) the circuit does not contain cut sets of either current sources or capacitors.

The format for a data statement consists of (1) the element name, (2) the circuit nodes to which the element is connected, and (3) the values of the parameters that describe the behavior of the element. Remember that the element name *must* start with a letter of the alphabet. For passive circuit elements and independent sources, the third part of the data statement is the numerical value of the circuit element or source. For dependent sources, the third part of the data statement contains the controlling nodes and the gain.

Before concluding our introduction to data statements, we need to pause and comment on polarity references. In MicroSim PSpice, when polarity is relevant to the behavior of the element, the first node is positive with respect to the second node. This implies that the current reference direction is from the first-named node to the second-named node.

CONTROL STATEMENTS

As we mentioned earlier, MicroSim PSpice was developed to analyze circuits containing integrated-circuit devices. One of the important aspects of such analysis is to determine the dc operating point, or bias level, of the circuit. As a result, MicroSim PSpice will automatically perform a dc analysis prior to both transient and ac analysis. Therefore it is possible to get a dc solution by just describing the circuit.

When you want other than simple dc analysis, you must use a control statement. The control statement consists of a MicroSim PSpice command followed by fields that describe the parameters of the desired analysis. The number and type of these parameter fields depend on the type of analysis that MicroSim PSpice performs. The analysis commands that are used in the study of linear circuits include .DC for complete dc analysis, .AC for ac analysis (frequency response), .TRAN for transient analysis (time response), .FOUR for Fourier analysis, .OP to compute the dc operating point of a circuit prior to other analysis, and .TF to compute the gain and input and output resistance (the transfer function). There is a summary of these commands at the end of this chapter.

OUTPUT STATEMENTS

MicroSim PSpice output has several constituent parts. The presence of each of these parts and the format used are determined by the output statements in the source file. These statements are briefly described below, illustrated in the examples, and summarized at the end of this chapter.

The first section of the output file is the description of the circuit itself. At a minimum, this includes a listing of the source file that you created for MicroSim PSpice to use in the analysis. The second section of the output file contains the default output from some of the analysis commands that generate output without any specific directions from the source file. For example, the .TF command always prints the gain, the input resistance, and the output resistance in the output file after analysis is complete.

The third section of the output file contains printout and line printer plots that have been explicitly requested in the source file. The .PRINT command prints data from the analysis. This statement includes fields that specify the type of analysis that generated the data and a list of the circuit variables whose values are to be printed. The .PLOT command creates line printer plots of data generated during MicroSim PSpice analysis. The plot statement includes the same fields as the print statement.

The fourth section contains information that summarizes some statistical information describing the execution of MicroSim PSpice.

This information includes the memory requirements and the execution time, and you can use it in conjunction with the specification of some optional parameters to make your analysis more efficient.

In addition, the MicroSim PSpice software is packaged with a graphics post-processor called MicroSim Probe. MicroSim Probe permits you to plot virtually any value of interest after MicroSim PSpice has analyzed the circuit you have described. MicroSim Probe needs access to a special data file in order to function properly. You can generate this data file during MicroSim PSpice analysis by including the .PROBE command in your source file. Once MicroSim PSpice has completed an error-free analysis of your circuit, you can activate MicroSim Probe either at the operating systems level or from the Command Shell. MicroSim Probe is illustrated in several of the examples.

We explore the MicroSim PSpice language and its use in analyzing linear circuits in the examples that follow. The chapter concludes with a brief summary of MicroSim PSpice commands used in linear circuit analysis.

EXAMPLE 1.1

a) Use MicroSim PSpice to find the voltages v_a and v_b in the circuit in Fig. 1.1. This circuit is similar to those introduced in Chapters 2, 3, and 4 in the text.

b) Use the MicroSim PSpice solutions to calculate (1) the total power dissipated in the circuit, (2) the power supplied by the independent current source, and (3) the power supplied by the current-controlled voltage source.

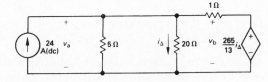

FIGURE 1.1 The circuit for Example 1.1.

SOLUTION

a) The circuit of Fig. 1.1 is redrawn in Fig. 1.2. We have assigned node numbers to all of the nodes in the circuit, assigning the number zero to the node chosen as the reference node. Note also the insertion of a zero-value dc voltage source in series with the 20-Ω resistor. We inserted this voltage source to act as an ammeter, so that we can print the value of the current i_Δ. The names of the circuit components are given in the figure to facilitate reading the MicroSim PSpice source file:

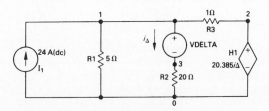

FIGURE 1.2 The circuit shown in Fig. 1.1 redrawn for MicroSim PSpice analysis.

```
Circuit Used to Check Power Dissipation
I1        0 1    DC        24
Vdelta    1 3    DC        0
H1        2 0    Vdelta    20.3846
R1        1 0    5
R2        3 0    20
R3        1 2    1
.END
```

This source file is in the file `\pspice\ex1.cir` on the disk. Since we require only a simple dc analysis, we need only describe the circuit components in the source file and do not need additional control or output statements.

The pertinent MicroSim PSpice output is

NODE	VOLTAGE	NODE	VOLTAGE	NODE	VOLTAGE	NODE	VOLTAGE
(1)	104.0000	(2)	106.0000	(3)	104.0000		

Hence

$$v_a = V(1) = 104.00 \text{ V}$$

and

$$v_b = V(2) = 106.00 \text{ V}.$$

b) 1. $P_{5\Omega} = 104^2/5 = 2163.20 \text{ W};$

$\quad\quad P_{20\Omega} = 104^2/20 = 540.80 \text{ W};$

$\quad\quad P_{1\Omega} = (106 - 104)^2/1 = 4.00 \text{ W}.$

$\quad\quad \sum P_{dis} = 2163.20 + 540.80 + 4.00 = 2708 \text{ W}.$

2. $P_{24A}(\text{supplied}) = 104(24) = 2496 \text{ W}.$

3. $P_{i\Delta}(\text{supplied}) = 106(2) = 212 \text{ W}.$

Note that the sum of power dissipated equals the sum of power supplied. Also note that the total-power-dissipated value given by MicroSim PSpice is 2500 W, which, when roundoff is taken into consideration, corresponds to the power supplied by the independent current source.

EXAMPLE 1.2

Use MicroSim PSpice to find the Thévenin equivalent with respect to terminals a,b for the circuit in Fig. 1.3. Thévenin equivalents are discussed in Chapter 4 of the text.

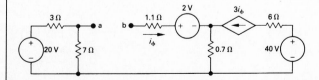

FIGURE 1.3 The circuit for Example 1.2.

SOLUTION

From the circuit shown in Fig. 1.3 note that i_ϕ is the current in an independent voltage source. Therefore we do not have to use a zero-value voltage source to measure the controlling current i_ϕ. Also note that there is only one connection to node b. MicroSim PSpice requires at least two connections to every node, so we must modify the circuit accordingly. We may do so in one of two ways. First, we may connect a resistor that is large compared to the other resistors in the circuit— say 10^6 Ω for this circuit—between node b and any other node in the circuit without affecting the behavior of the circuit. Second, we may connect a capacitor between node b and any other node. The capacitor behaves like an open circuit during dc analysis and therefore does not influence the dc Thévenin equivalent.

The circuit of Fig. 1.3 is redrawn in Fig. 1.4 in preparation for writing the MicroSim PSpice source file. Note that we connected a resistor of 10^6 Ω between nodes 0 and 3. Also note that we chose node b as the reference node. We did so because MicroSim PSpice automatically prints all the node-to-reference voltages, and therefore the Thévenin voltage becomes part of the printout if node b is the reference.

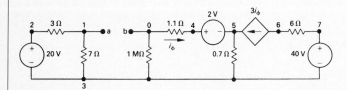

FIGURE 1.4 The circuit shown in Fig. 1.3 redrawn for MicroSim PSpice analysis.

With three independent sources, we may choose which one to use in the `.TF` control statement. As we are interested only in the Thévenin voltage and resistance with respect to terminals a,b, the choice is immaterial. In other words, the transfer function ratio and the input resistance at the source are not relevant to the Thévenin equivalent at a,b. In the MicroSim PSpice source file shown below, we selected the 20 V source as the input source:

```
Finding the Thevenin Equivalent when a Dependent Source is Present
R1       1 2      3
R2       1 3      7
V1       2 3      DC       20
R3       0 4      1.1
R4       0 3      1e6
V2       4 5      DC       2
R5       5 3      0.7
F1       6 5      V2       3
R6       6 7      6
V3       7 3      DC       40
.TF      V(1,0)   V1
.END
```

This source file is in the file `\pspice\ex2.cir` on the disk.
The relevant output from the analysis is

```
****      SMALL-SIGNAL CHARACTERISTICS
V(1, 0)/V1 = 7.000E-01
INPUT RESISTANCE AT V1 = 1.000E+01
OUTPUT RESISTANCE AT V(1,0) = 6.000E+00
```

Hence the Thévenin equivalent is a 14 V source in series with a 6 Ω resistor, as shown in Fig. 1.5.

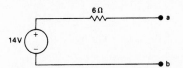

FIGURE 1.5 The Thévenin equivalent for Example 1.2.

E X A M P L E   1.3

The current source in the circuit shown in Fig. 1.6 varies from 0 to 5 A in 1 A steps. For each value of the current, tabulate v_o and i_g as v_g varies from 0 to 100 V in steps of 20 V. The text describes circuit analysis techniques for circuits with two or more independent dc sources in Chapter 4.

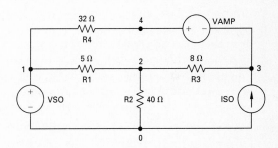

FIGURE 1.6 The circuit for Example 1.3.

S O L U T I O N

Figure 1.7 shows the circuit prepared for MicroSim PSpice analysis. In order to print out the current i_o, we insert a zero-value voltage source in series with the 32 Ω resistor. We also need to specify a dc analysis where the voltage and current are stepped through a range of values, using the `.DC` control statement. Finally, we include a `.PRINT` statement to identify the names of the voltage and current variables whose values will be printed at each step. The MicroSim PSpice source file is

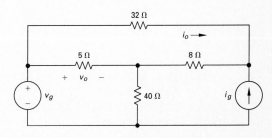

FIGURE 1.7 The circuit in Fig. 1.6 prepared for MicroSim PSpice analysis.

```
Circuit with both Voltage and Current Sources Varied
Vso       1 0     DC      0
Iso       5 3     DC      0
Vamp      0 5     DC      0
R1        1 2     5
R2        2 0     40
R3        2 3     8
R4        1 3     32
.DC     Vso       0       100     20      Iso     0       5       1
.PRINT  DC      V(1,2)  I(amp)
.END
```

This source file is in the file \pspice\ex3.cir on the disk.

The output from the analysis is:

Vso	V(1,2)	I(amp)
0.000E+00	0.000E+00	0.000E+00
2.000E+01	2.000E+00	-1.840E-11
4.000E+01	4.000E+00	-3.680E-11
6.000E+01	6.000E+00	-5.520E-11
8.000E+01	8.000E+00	-7.360E-11
1.000E+02	1.000E+01	-9.200E-11
0.000E+00	-3.200E+00	1.000E+00
2.000E+01	-1.200E+00	1.000E+00
4.000E+01	8.000E-01	1.000E+00
6.000E+01	2.800E+00	1.000E+00
8.000E+01	4.800E+00	1.000E+00
1.000E+02	6.800E+00	1.000E+00
0.000E+00	-6.400E+00	2.000E+00
2.000E+01	-4.400E+00	2.000E+00
4.000E+01	-2.400E+00	2.000E+00
6.000E+01	-4.000E-01	2.000E+00
8.000E+01	1.600E+00	2.000E+00
1.000E+02	3.600E+00	2.000E+00
0.000E+00	-9.600E+00	3.000E+00
2.000E+01	-7.600E+00	3.000E+00
4.000E+01	-5.600E+00	3.000E+00
6.000E+01	-3.600E+00	3.000E+00
8.000E+01	-1.600E+00	3.000E+00
1.000E+02	4.000E-01	3.000E+00
0.000E+00	-1.280E+01	4.000E+00
2.000E+01	-1.080E+01	4.000E+00
4.000E+01	-8.800E+00	4.000E+00
6.000E+01	-6.800E+00	4.000E+00
8.000E+01	-4.800E+00	4.000E+00
1.000E+02	-2.800E+00	4.000E+00
0.000E+00	-1.600E+01	5.000E+00
2.000E+01	-1.400E+01	5.000E+00
4.000E+01	-1.200E+00	5.000E+00
6.000E+01	-1.000E+01	5.000E+00
8.000E+01	-8.000E+00	5.000E+00
1.000E+02	-6.000E+00	5.000E+00

This example illustrates the use of MicroSim PSpice in analyzing the behavior of a circuit for a wide variety of conditions, which would be tedious to calculate by hand.

EXAMPLE 1.4

Let the input to the two-stage op amp amplifier circuit shown in Fig. 1.8 be a dc voltage with a magnitude of 0.5 V. Use MicroSim PSpice to calculate the voltage at the output of each stage.

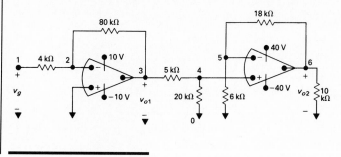

FIGURE 1.8 The two-stage amplifier circuit for Example 1.4.

SOLUTION

MicroSim PSpice offers three options for describing an operational amplifier (op amp) within a source file. The first option is to model the op amp using resistors and dependent sources. Such a model is described in Section 5.6 of the text. The second option is similar to the first option, but now the model is made into a MicroSim PSpice subcircuit and assigned a unique name. Once the op amp subcircuit exists, you can use it like any other MicroSim PSpice circuit element, and include it in multiple locations within the circuit being described. The third option is to take advantage of op amp models already supplied with MicroSim PSpice. These models are available in a device library that you can access with the .LIB command. The library models are considerably more sophisticated than the simple dependent-source-based models, and are designed to mimic the characteristics of actual op amps.

In general, if you have only one op amp in your circuit, it is best to model it using resistors and a dependent source. If the circuit contains two or more op amps, it is a good idea to create an op amp subcircuit which you can use to represent each op amp in the circuit. You should only use the library op amp models when it is important to the circuit analysis that certain characteristics of the op amp be modeled accurately. Because the library models are complex, circuits containing them tend to occupy more space, and analysis of such circuits takes longer.

In this example, we use a simple subcircuit model of an op amp. The subcircuit model is taken from Fig. 5.15 in the text and is shown here in Fig. 1.9. We assume that the op amps in the circuit in Fig. 1.8 are identical, so once we have defined and named the op amp subcircuit

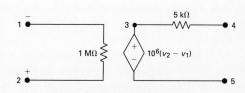

FIGURE 1.9 An op amp subcircuit model.

model, we can use it just like any other component in describing the two-stage amplifier to MicroSim PSpice. The MicroSim PSpice source file is

```
PSpice Source File with a Repeated Subcircuit
Vg         1 0      DC        0
Rs1        1 2      4e3
Rf1        2 3      80e3
R1         3 4      5e3
R2         4 0      20e3
Rs2        5 0      6e3
Rf2        5 6      18e3
RL         6 0      10e3
.SUBCKT  IDEAL_OPAMP      1 2 4 5
R1         2 1      1e6
E1         3 5 2 1 1e6
RO         3 4      5e3
.ENDS    IDEAL_OPAMP
X1         2 0 3 0 IDEAL_OPAMP
X2         5 4 6 0 IDEAL_OPAMP
.DC       Vg       0.5      0.5       1
.PRINT   DC        V(3)     V(6)
.END
```

This source file is in the file \pspice\ex4.cir on the disk. Note the use of the .DC control statement so that only the node voltages of interest are printed out. For convenience we have used the same node designations in the OPAMP subcircuit as used in Fig. 1.9.

The outputs from the MicroSim PSpice program are

```
Vg            V(3)          V(6)
5.000E-01   -1.000E+01   -3.200E+01
```

Therefore,

$$v_{o1} = V(3) = -10 \text{ V}$$

and

$$v_{o2} = V(6) = 32 \text{ V}.$$

Analysis of the two-stage amplifier circuit shows that these are the results expected when the op amps are assumed to be ideal. The values chosen for the components in the op amp subcircuit model in Fig. 1.9 make the op amps very nearly ideal. Try changing the component values in the op amp subcircuit model and note their effect on the MicroSim PSpice analysis. You might also want to replace the subcircuit models for the op amps with more realistic models from the op amp device library and note the effect on the amplifier gain.

EXAMPLE 1.5

Use MicroSim PSpice to examine the effect of the resistor value on the natural response of the series *RLC* circuit shown in Fig. 1.10. Vary the value of the resistor from 20 Ω to 100 Ω in 20 Ω steps, and use MicroSim Probe to display the value of v_c versus t for each of the resistor values. The natural response of the series *RLC* circuit is explored in Chapter 8 of the text.

SOLUTION

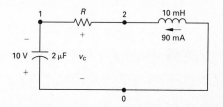

FIGURE 1.10 The series *RLC* circuit for Example 1.5.

You can find the MicroSim PSpice source file in the file \pspice\ex5.cir on the disk. It is listed below:

```
Effect of Varying R on RLC Natural Response
C1       1 0     2e-6
L2       2 0     10e-3   IC = -90e-3
R1       1 2     RMOD    1
.IC      V(1) = -10
.MODEL   RMOD    RES(R=1)
.STEP    LIN     RES     RMOD(R)           20,100,20
.TRAN    20e-6   2000e-6         UIC
.PROBE
.END
```

There are several important things to note. First, we have specified the resistor value using a model name, and then defined the resistor model using the .MODEL statement. The model statement includes the parameter RES, which is stepped through the values of interest using the control statement .STEP. In this way, we can have MicroSim PSpice analyze the series *RLC* circuit many times, for each value of resistance, without having to specify additional circuits to MicroSim PSpice.

Next, note that the control statement .TRAN is used to request that MicroSim PSpice perform transient analysis, meaning that it will calculate the values of voltage and current as time is varied. The two numbers specified in the .TRAN statement specify the time increment (20 μs) and the final value of the time (2000 μs). You need to perform a preliminary analysis on the circuit in order to compute the time constant of the circuit, which will enable you to choose the time increment and final time specified to MicroSim PSpice. Including UIC in the .TRAN statement causes MicroSim PSpice to use the initial condition specified in the inductor data statement, rather than computing the initial conditions from the circuit description. Finally, note that the output statement .PROBE causes MicroSim PSpice to write data to a file that the graphics post-processor program MicroSim Probe uses to produce plots of the analysis results.

Figure 1.11 shows the MicroSim Probe plot of the capacitor voltage versus time. Note that the analysis using the smallest value of resistance, denoted □ on the graph, has the smallest damping ratio. Increasing the value of resistance increases the damping ratio. The largest value of resistance is denoted ▽ on the graph. MicroSim PSpice and MicroSim Probe provide an easy way to visualize the results of time-consuming circuit analysis.

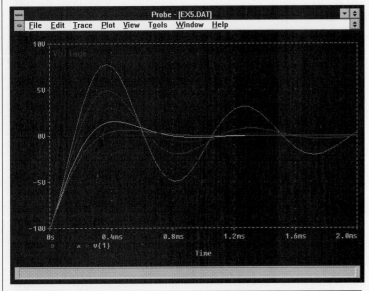

FIGURE 1.11 The plot of the capacitor voltage versus time as the resistance is varied in the series *RLC* circuit of Fig. 1.10.

EXAMPLE   1.6

Use MicroSim PSpice and MicroSim Probe to examine the effect of the capacitor value on the frequency response of the parallel *RLC* circuit shown in Fig. 1.12. Step the capacitor values from 0.15 μF through 0.35 μF in increments of 0.05 μF. Plot your results and comment on the effect of the capacitor value on the frequency response. The text analyzes the frequency response of a parallel *RLC* circuit in Chapter 15.

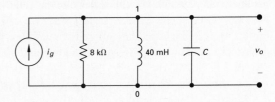

FIGURE 1.12 The parallel *RLC* circuit for Example 1.6.

SOLUTION

You can find the MicroSim PSpice source file in the file \pspice \ex6.cir on the disk. It is listed below:

```
Effects of Varying C on Parallel RLC Frequency Response
Ig       0 1  AC       50e-3    0
R1       1 0      8e3
L1       1 0      40e-3
C1       1 0      CMOD     1
.MODEL   CMOD     CAP(C=1)
.STEP    CAP      CMOD(C)            0.15e-6  0.35e-6  0.05e-6
.AC      LIN      101      500      2500
.PROBE
.END
```

Note that the .MODEL and .STEP statements vary the capacitor value in the circuit through the desired range of values. This time we use the .AC command statement to specify ac steady-state analysis using a range of values for the source frequency. Computing the center frequency for the parallel *RLC* circuit will be useful in deciding on a range of frequency values. Here we specify a starting value of 500 Hz and a final value of 2500 Hz, and MicroSim PSpice analyzes the circuit for 101 frequency values equally spaced in the linear range between these two end points.

Figure 1.13 shows the MicroSim Probe plot for the output voltage and each of the five capacitor values. The smallest value of capacitance produced the plot farthest to the right, denoted □. As the capacitance increases, the plots move to the left. Hence the resonant frequency *decreases* as the capacitance *increases*. We expect this result because the equation for resonant frequency for a parallel *RLC* circuit is

$$\omega_r = \sqrt{\frac{1}{LC}}.$$

Also, as the capacitance increases, the resonant peak becomes sharper. That is, as the capacitance increases, the quality becomes higher. This result, too, comes as no surprise, because the equation for *Q* in a parallel *RLC* circuit is

$$Q = R\sqrt{\frac{C}{L}}.$$

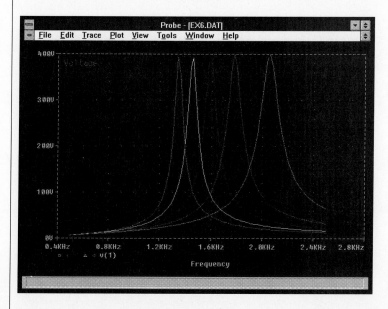

FIGURE 1.13 The relationship between capacitor value and the frequency response of the parallel *RLC* circuit.

A QUICK REFERENCE TO MICROSIM PSPICE

DATA STATEMENTS

In the syntax for data statements, xxx represents any alphanumeric string used to identify a circuit component, n+ represents the positive voltage node, and n− represents the negative voltage node. Positive voltage drop is from n+ to n−, and positive current flows from n+ to n−. Any element in the syntax specification that is optional is enclosed by brackets ([ ]).

INDEPENDENT DC VOLTAGE AND CURRENT SOURCES

SYNTAX (VOLTAGE)   Vxxx   n+   n−   DC   value

SYNTAX (CURRENT)   Ixxx   n+   n−   DC   value

where

value is the dc voltage in volts for the voltage source and the dc current in amperes for the current source.

DESCRIPTION   Provides a constant source of voltage or current to the circuit.

EXAMPLES   Vin   1   2   DC   3.5
              Isource   3   0   DC   0.25

INDEPENDENT AC VOLTAGE AND CURRENT SOURCES

SYNTAX (VOLTAGE)   Vxxx   n+   n−   AC   mag   [phase]

SYNTAX (CURRENT)   Ixxx   n+   n−   AC   mag   [phase]

where

mag is the magnitude of the ac waveform in volts for the voltage source and in amperes for the current source;

phase is the phase angle of the ac waveform in degrees and has a default value of 0.

DESCRIPTION   Supplies a sinusoidal voltage or current at a fixed frequency to the circuit.

EXAMPLES   Vab   3   2   AC   10   90
              I12   4   6   AC   0.5

INDEPENDENT TRANSIENT VOLTAGE AND CURRENT SOURCES

SYNTAX (VOLTAGE)   `Vxxx    n+    n-    trans_type`

SYNTAX (CURRENT)   `Ixxx    n+    n-    trans_type`

where `trans_type` is one of the following transient waveform types:

* *Exponential*
  `EXP(start peak [delay1] [rise] [delay2] [fall])`

  where

  `start` is the initial value of the voltage in volts or current in amperes;

  `peak` is the maximum value of the voltage in volts or current in amperes;

  `delay1` is the delay in seconds prior to a change in voltage or current from the initial value, and is 0 by default;

  `rise` is the time constant in seconds of the exponential decay from start to peak, and has a default value of `tstep` seconds (see `.TRAN`);

  `delay2` is the delay in seconds prior to introducing a second exponential decay, from the maximum value toward the initial value, and has a default value of `delay1 + tstep` seconds (see `.TRAN`);

  `fall` is the time constant in seconds of the exponential decay from peak to final, and has a default value of `tstep` seconds (see `.TRAN`).

* *Pulsed*
  `PULSE(min max [delay] [rise] [fall] [width] [period])`

  where

  `min` is the minimum value of the waveform in volts for the voltage source and in amperes for the current source;

  `max` is the maximum value of the waveform in volts for the voltage source and in amperes for the current source;

  `delay` is the time in seconds prior to the onset of the pulse train, and has a default of 0 seconds;

  `rise` is the time in seconds for the waveform to transition from `min` to `max`, and has a default of `tstep` seconds (see `.TRAN`);

fall is the time in seconds for the waveform to transition from max to min, and has a default value of tstep seconds (see .TRAN);

width is the time in seconds that the waveform remains at the maximum value, and has a default of tstop seconds (see .TRAN);

period is the time in seconds that separates the pulses in the pulse train, and has a default of tstop seconds (see .TRAN).

* *Piecewise linear*

   PWL(t1 val1 t2 val2 ...)

   where

   t1 val1 are a paired time (in seconds) and value (in volts for a voltage source and amperes for a current source) that specify a corner of the waveform; all pairs of times and values are linearly connected to form the whole waveform.

* *Damped sinusoid*

   SIN(off peak [freq] [delay] [damp] [phase])
   where

   off is the initial value of the voltage in volts or the current in amperes;

   peak is the maximum amplitude of the voltage in volts or the current in amperes;

   freq is the sinusoidal frequency in hertz, and has a default value of 1/tstop hertz (see .TRAN);

   delay is the time in seconds that the waveform stays at the initial value before the sinusoidal oscillation begins, and has a default value of 0 seconds;

   damp is the sinusoidal damping factor in seconds^{-1} used to specify the decaying exponential envelope for the sinusoid, and has a default value of 0;

   phase is the initial phase angle of the sinusoidal waveform in degrees, and has a default value of 0.

DESCRIPTION   Supplies a time-varying voltage or current to the circuit whose waveform can be characterized as exponential, pulsed, piecewise linear, or damped sinusoidal.

EXAMPLES   Vs   3   1   EXP(2   6   .5   .1   .5   .2)
                 Iin 4   5   PULSE(-.3   .3   0   .01   .01   1   2)
                 I5   2   3   PWL(0   .2   1   .6   1.5   .6   3   -.5)
                 Vg   0   3   SIN(0   2   100   0   0   90)

DEPENDENT VOLTAGE-CONTROLLED VOLTAGE SOURCE

SYNTAX   `Exxx  n+  n-  cn+  cn-  gain`

where

`cn+` is the positive node for the controlling voltage;

`cn-` is the negative node for the controlling voltage;

`gain` is the ratio of the source voltage (between `n+` and `n-`) to the controlling voltage (between `cn+` and `cn-`).

DESCRIPTION   Provides a voltage source whose value depends on a voltage measured elsewhere in the circuit.

EXAMPLE   `Eop  1  2  4  0  .5`

DEPENDENT CURRENT-CONTROLLED CURRENT SOURCE

SYNTAX   `Fxxx  n+  n-  Vyyy  gain`

where

`Vyyy` is the name of the voltage source through which the controlling current flows;

`gain` is the ratio of the source current (flowing from `n+` to `n-`) to the controlling current (flowing through `Vyyy`).

DESCRIPTION   Provides a current source whose value depends on the magnitude of a current flowing through a voltage source elsewhere in the circuit.

EXAMPLE   `Fdep  3  2  Vcontrol  10`

DEPENDENT VOLTAGE-CONTROLLED CURRENT SOURCE

SYNTAX   `Gxxx  n+  n-  cn+  cn-  gain`

where

`cn+` is the positive node for the controlling voltage;

`cn-` is the negative node for the controlling voltage;

`gain` is the ratio of the source current (flowing from `n+` to `n-`) to the controlling voltage (between `cn+` and `cn-`), in siemens.

DESCRIPTION   Provides a current source whose value depends on the magnitude of a voltage measured elsewhere in the circuit.

EXAMPLE   `Gon  3  6  4  1  0.35`

DEPENDENT CURRENT-CONTROLLED VOLTAGE SOURCE

SYNTAX   `Hxxx  n+  n-  Vyyy  gain`

where

`Vyyy` is the name of the voltage source through which the controlling
current flows;

`gain` is the ratio of the source voltage (between `n+` and `n-`) to the
controlling current (flowing through `Vyyy`), in ohms.

DESCRIPTION   Provides a voltage source whose value depends on the
magnitude of the current flowing through another voltage source
elsewhere in the circuit.

EXAMPLE   `Hout  7  2  Vdummy  -2.5e-3`

VOLTAGE-CONTROLLED SWITCH

SYNTAX   `Sxxx  n+  n-  cn+  cn-  mname`

where

`cn+` is the positive node for the controlling voltage;

`cn-` is the negative node for the controlling voltage;

`mname` is the name of a `VSWITCH` model defined in a `.MODEL`
statement (see `.MODEL`).

DESCRIPTION   Simulates a switch that is opened or closed, depending on
the magnitude of a voltage measured elsewhere in the circuit.

EXAMPLES   `S32  13  12  4  0  smod`
          `.MODEL  smod  VSWITCH(RON = 0.1)`

CURRENT-CONTROLLED SWITCH

SYNTAX   `Sxxx  n+  n-  Vyyy  mname`

where

`Vyyy` is the name of the voltage source through which the controlling
current flows;

`mname` is the name of an `ISWITCH` model defined in a `.MODEL`
statement (see `.MODEL`).

DESCRIPTION   Simulates a switch that is opened or closed, depending
on the magnitude of a current flowing through a voltage source
located elsewhere in the circuit.

EXAMPLES   `Woff  4  8  Vab  imod`
`.MODEL  imod  ISWITCH(ROFF = 1e+3, ION = 1e-6)`

RESISTOR

SYNTAX   `Rxxx  n+  n-  [mname]  value`

where

`mname` is the name of an `RES` model defined in a `.MODEL` statement
(see `.MODEL`)—note that `mname` is optional;

`value` is the resistance, in ohms.

DESCRIPTION   Models a resistor, a circuit element whose voltage and
current are linearly dependent.

EXAMPLE   `Rfor   3   4   16e+3`

INDUCTOR

SYNTAX   `Lxxx n+ n- [mname] value [IC = icval]`

where

`mname` is the name of an `IND` model defined in a `.MODEL` statement
(see `.MODEL`)—note that `mname` is optional;

`value` is the inductance, in henries;

`icval` is the initial value of the current in the inductor, in amperes—
note that specifying the initial current is optional.

DESCRIPTION   Models an inductor, a circuit element whose voltage is
linearly dependent on the derivative of its current.

EXAMPLE   `L44   1   9   3e-3   IC = 1e-2`

CAPACITOR

SYNTAX   `Cxxx n+ n- [mname] value [IC = icval]`

where

`mname` is the name of a `CAP` model defined in a `.MODEL` statement
(see `.MODEL`)—note that `mname` is optional;

`value` is the capacitance, in farads;

`icval` is the initial value of the voltage across the capacitor, in
volts—note that specifying the initial voltage is optional.

DESCRIPTION   Models a capacitor, a circuit element whose current is linearly dependent on the derivative of its voltage.

EXAMPLES   `Ctwo   4   0   cmod   2e-6`
`.MODEL   cmod   CAP(C=1)`

MUTUAL INDUCTANCE

SYNTAX   `Kxxx Lyyy Lzzz value`

where

`Lyyy` is the name of the inductor on the primary side of the coil;

`Lzzz` is the name of the inductor on the secondary side of the coil;

`value` is the mutual coupling coefficient k, which has a value such that $0 \leq k \leq 1$.

DESCRIPTION   Models the magnetic coupling between any two inductor coils in a circuit.

EXAMPLE   `Kab   La   Lb   0.5`

SUBCIRCUIT DEFINITION

SYNTAX   `.SUBCKT   name   [nodes]   .ENDS`

where

`name` is the name of the subcircuit, as referenced by an X statement (see Subcircuit Call);

`nodes` is the optional list of nodes used to identify the connections to the subcircuit;

`.ENDS` signifies the end of the subcircuit definition.

DESCRIPTION   Used to provide "macro"-type definitions of portions of a circuit. When the subcircuit is referenced in an X statement, the definition between the `.SUBCKT` statement and the `.ENDS` statement replaces the X statement in the source file.

EXAMPLE   `.SUBCKT   opamp   1   2   3   4   5`
`   ⋮`
`.ENDS`

SUBCIRCUIT CALL

SYNTAX Xxxx   [nodes]   name

where

nodes is the optional list of circuit nodes used to connect the sub-circuit into the rest of the circuit; there must be as many nodes in this list as there are in the subcircuit definition (see Subcircuit Definition);

name is the name of the subcircuit as defined in a .SUBCKT state-ment.

DESCRIPTION Replaces the X statement with the definition of a subcir-cuit, which permits a subcircuit to be defined once and used many times within a given source file.

EXAMPLE Xamp 4 2 8 5 1 opamp

LIBRARY FILE

SYNTAX .LIB [fname]

where

fname is the name of the library file containing .MODEL or .SUBCKT statements referenced in the source file, and by de-fault is the nominal or evaluation library file.

DESCRIPTION Used to reference a library of models or subcircuits.

EXAMPLE .LIB mylib.lib

DEVICE MODELS

SYNTAX .MODEL mname mtype [(par = value)]

where

mname is a unique model name, which is also used in the device statement that incorporates this model;

mtype is one of the model types available;

par = value is an optionally specified list of parameters and their assigned values, specific to the model type:

- VSWITCH, which models a voltage-controlled switch and has the following parameters:

  RON, the resistance of the closed switch, whose default value is 1 Ω;

ROFF, the resistance of the open switch, whose default value is 10^6 Ω;

VON, the control voltage level necessary to close the switch, whose default value is 1 V;

VOFF, the control voltage level necessary to open the switch, whose default value is 0 V;

* ISWITCH, which models a current-controlled switch and has the following parameters:

RON, the resistance of the closed switch, whose default value is 1 Ω;

ROFF, the resistance of the open switch, whose default value is 10^6 Ω;

VON, the control current necessary to close the switch, whose default value is 10^{-3} A;

VOFF, the control current necessary to open the switch, whose default value is 0 A;

* RES, which models a resistor and has the parameter R, the resistance multiplier, whose default value is 1;

* IND, which models an inductor and has the parameter L, the inductance multiplier, whose default value is 1;

* CAP, which models a capacitor and has the parameter C, the capacitance multiplier, whose default value is 1.

DESCRIPTION   Defines standard devices that can be used in a circuit and sets parameter values that characterize the specific device being modeled.

EXAMPLES
```
.MODEL    lmodel    IND(L=2)
.MODEL    resist    RES
```

CONTROL STATEMENTS

DC ANALYSIS

SYNTAX   `.DC [type] vname start end incre [nest-sweep]`

where

type is the sweep type and must be LIN for a linear sweep from the start value to the end value, OCT for a logarithmic sweep from start value to end value in octaves, DEC for a logarithmic sweep from start value to end value in decades, or LIST if a list of values is to be used; the default sweep type is LIN;

vname is the name of the circuit element whose value is swept, usually an independent voltage or current source;

start is the beginning value of vname;

end is the final value of vname; note that when the sweep type is LIST, start, end, and incre are replaced by a list of values that vname will take on;

incre is the step size for the LIN sweep type and is the number of points per octave or decade for the OCT and DEC sweep types;

nest-sweep is an optional additional sweep of a second circuit variable, which follows the same syntax as the sweep for the first variable.

DESCRIPTION   Provides a method for varying one or two dc source values and analyzing the circuit for each sweep value.

EXAMPLES   `.DC  Vab  1   10   .5`
`.DC  I1  DEC  .1  10  30  V1  DEC  10  100 30`

TRANSFER FUNCTION

SYNTAX   `.TF  output   input`

where

output is the output circuit variable, which is a voltage or a current;

input is the source circuit variable, which is a voltage or a current.

DESCRIPTION   Computes the gain, input resistance, and output resistance between source and output.

EXAMPLE   `.TF  V(6,2)  Vsource`

SENSITIVITY

SYNTAX   `.SENS vname`

where

vname is the circuit variable name (or a list of circuit variable names) for which sensitivity analysis will be performed.

DESCRIPTION   Computes and prints to the output file a dc sensitivity analysis of `vname` to the values of other circuit elements, including resistors, independent voltage and current sources, and voltage- and current-controlled switches.

EXAMPLE   `.SENS  V(2)  I(Vin)`

TRANSIENT ANALYSIS

SYNTAX   `.TRAN tstep tstop [npval] istep [UIC]`

where

`tstep` is the time interval separating values that are printed or plotted;

`tstop` is the ending time for the transient analysis;

`npval` is the time between 0 and the first value printed or plotted, which by default is 0;

`istep` is the internal time step used for computing values, which by default is `tstop`/50;

`UIC` will bypass the calculation of the bias point, which usually precedes the transient analysis, and use the initial conditions specified by `IC =` in inductor and capacitor data statements instead.

DESCRIPTION   Performs a transient analysis of the circuit described in the source file to calculate the values of circuit variables as a function of time.

EXAMPLE   `.TRAN  .01  10  0  .001  UIC`

INITIAL CONDITION

SYNTAX   `.IC  Vnode = value`

where

`Vnode = value` is one or a list of pairs consisting of a voltage node `Vnode`, represented in standard form, and the initial value in volts at that node.

DESCRIPTION   Sets the initial conditions for transient analysis by specifying one or more node voltage values.

EXAMPLE   `.IC  V(1) = 10  V(3,4) = -2`

AC ANALYSIS

SYNTAX   `.AC [type] num start end`

where

`type` is the type of sweep, which must be one of the following keywords:

> `LIN` for a linear sweep in frequency, which is the default;
>
> `OCT` for a logarithmic sweep in frequency by octaves;
>
> `DEC` for a logarithmic sweep in frequency by decades;

`num` is the total number of points in the sweep for a linear sweep and the number of points per octave or decade for a logarithmic sweep;

`start` is the starting frequency, in hertz;

`end` is the final frequency, in hertz.

DESCRIPTION   Computes the frequency response of the circuit described in the source file as the frequency is swept either linearly or logarithmically from an initial value to a final value.

EXAMPLE   `.AC LIN 300 10 1000`

STEPPED VALUES

SYNTAX   `.STEP [type] name start end incre`

where

`type` is one of the following keywords describing the type of parameter variation:

> `LIN` for a linear variation, which is the default;
>
> `OCT` for a logarithmic variation by octaves;
>
> `DEC` for a logarithmic variation by decades;

`name` is the name of the circuit element whose value is to be varied;

`start` is the starting value for the circuit element, in units appropriate to the type of circuit element;

`end` is the final value for the circuit element, in units appropriate to the type of circuit element;

incre is the step size for a linear variation and the number of values per octave or decade for the logarithmic variation. Note that when you want a discrete list of parameter values, type is not used and name is followed by the keyword LIST and a list of parameter values.

DESCRIPTION   Used to step a circuit element's value through a range either linearly or logarithmically, or through a discrete list, analyzing the circuit for each value.

EXAMPLES   `.STEP  DEC  RES  RMOD(R)  10  10e4  3`
`.STEP  V2  LIST  1  4  12  19`

FOURIER SERIES ANALYSIS

SYNTAX   `.FOUR  freq  vname`

where

freq is the fundamental frequency of the circuit;

vname is the variable name or the list of variable names for which Fourier series coefficients will be computed.

DESCRIPTION   Uses the results of a transient analysis to compute the Fourier coefficients for the first nine harmonics. Note that the .FOUR statement requires a .TRAN statement to perform the transient analysis.

EXAMPLE   `.FOUR  10e3  V(3)  V(6,2)`

OUTPUT STATEMENTS

OPERATING POINT

SYNTAX   `.OP`

DESCRIPTION   Outputs information describing the bias point computations for the circuit being simulated.

EXAMPLE   `.OP`

PRINT RESULTS

SYNTAX   `.PRINT  type  vname`

where

`type` identifies the type of analysis performed to generate the data being printed and must be one of the keywords DC, AC, or TRAN;

`vname` is the variable name or the list of variable names for which values are to be printed.

DESCRIPTION Prints the results of circuit analysis in table form to an output file for each circuit variable specified.

EXAMPLE `.PRINT AC V(3,2) I(Vsource)`

PLOT RESULTS

SYNTAX `.PLOT type vname [low] [high]`

where

`type` identifies the type of analysis used to produce the values being plotted and must be one of the keywords DC, AC, or TRAN;

`vname` is the variable name or the list of variable names for which values are to be plotted;

`low` and `high` optionally set the range of values of the independent variable to be plotted, which default to the entire range of values.

DESCRIPTION Generates line-printer plots of circuit variables computed during dc, ac, or transient analysis.

EXAMPLE `.PLOT TRANS I(Vin) V(4)`

PROBE

SYNTAX `.PROBE [vname]`

where

`vname` is the optionally specified variable or list of variables whose values from dc, ac, and transient analysis will be stored in the file that MicroSim Probe uses to generate plots. By default, values of all circuit variables will be stored in the MicroSim Probe file.

DESCRIPTION Generates a file of data from dc, ac, and transient analysis used by Probe to generate high-quality plots.

EXAMPLE `.PROBE V(2) V(3) V(4) I(Rin) I(Rout)`

MISCELLANEOUS STATEMENTS

TITLE LINE

DESCRIPTION   Each source file must have a title line as its first line.

EXAMPLE   `A Circuit to Simulate RLC`
`*    Frequency Response`

END STATEMENT

SYNTAX   `.END`

DESCRIPTION   Each source file must have an `.END` statement as its final line.

EXAMPLE   `.END`

PRINTING CONTROL

SYNTAX   `.WIDTH  OUT = value`

where

`value` is used to specify the number of columns used in printing the output file, which must be either 80 (the default) or 132.

DESCRIPTION   Used to change the number of columns in the output file from the default.

EXAMPLE   `.WIDTH  OUT = 132`

SIMULATION OPTIONS

SYNTAX   `.OPTIONS  oname  [oname = value]`

where

`oname` is the name or the list of names of option flags to be set;

`oname = value` is the name and assigned value or the list of names and their values for options requiring values.

DESCRIPTION   Used to override the default specifications of many parameters used to control circuit simulation.

EXAMPLE   `.OPTION  NOPAGE  NONODE  RELTOL = .1`

BIBLIOGRAPHY

Walter Banzhaf, *Computer-Aided Circuit Analysis Using PSpice*, second ed., Regents/Prentice Hall, Englewood Cliffs, NJ, 1992.

L. H. Fenical, *PSpice: A Tutorial*, Regents/Prentice Hall, Englewood Cliffs, NJ, 1992.

James G. Gottling, *Hands-On PSpice*, Houghton Mifflin Co., Boston, MA, 1995.

John Keown, *PSpice and Circuit Analysis*, second ed., Merrill/Macmillan Publishing Co., NY, 1993.

MicroSim Corporation, *The Design Center: Circuit Analysis — Reference Manual*, Irvine, CA, 1994.

MicroSim Corporation, *The Design Center: Circuit Analysis — User's Guide*, Irvine, CA, 1994.

Muhammad H. Rashid, *Spice for Circuits and Electronics Using PSpice*, second ed., Prentice Hall, Englewood Cliffs, NJ, 1995.

Paul W. Tuinenga, *A Guide to Circuit Simulation and Analysis Using PSpice*, second ed., Prentice Hall, Englewood Cliffs, NJ, 1992.

2 SCHEMATIC CAPTURE AND CIRCUIT ANALYSIS

Like the circuit simulation tools described in Chapter 1, tools that perform schematic capture and circuit analysis are strongly connected to the study of electric circuits. In general, a schematic capture software package permits you to draw the circuit to be simulated on the screen, mimicking the construction of an actual circuit on a breadboard. You create the circuit by selecting parts from a bin of available parts, assigning values to each of the parts, and wiring the circuit together. Once the circuit exists as a drawing, it is analyzed to produce a netlist, which is a listing of all elements in the circuit that identifies their topological relationship in the circuit. From the netlist, you can simulate a circuit. The simulation is performed by software programmed to solve the underlying mathematical equations that describe the circuit's behavior. Finally, you can get the results of the simulation either as a listing of current and voltage values or through interrogation using a graphing package. One other step in this process, which is not of interest here, involves using the netlist to create a printed circuit board layout of an actual circuit, in preparation for its manufacture. Therefore, this category of software tools enables you to start with a drawing of a circuit schematic, simulate and analyze the performance of your design, and once you are satisfied with the design, create a printed circuit board ready for components to be attached.

Schematic capture and circuit analysis tools have many important uses in the study of electric circuits. We draw your attention to three of these uses. Schematic capture and circuit simulation make the study of topologically complex circuits more tractable. The problems typically encountered in the study of linear electric circuits are topologically simple, and lead to a description of a circuit's behavior using one, two, or at most three equations. There are many interesting and practical circuits that you can describe with 6 or 10 or 15 simultaneous equations. The behavior of such circuits would be tedious to describe without computational support.

Schematic capture and circuit simulation permit you to study the effect of changing component values, or examine the influence of the tolerance of component values, on the behavior of a circuit with minimal effort. It is easy to specify and simulate a circuit where one component can take on a variety of different values. Using a graphical interface to examine the data from such a simulation will enable you to see the effect of changing that component value on the circuit variables of interest. This leads to the third important use of schematic capture and circuit analysis tools — they often include a graphical interface for examining results.

We use three tools from MicroSim to illustrate schematic capture and circuit simulation:[1]

- MicroSim Schematics, the program that allows you to draw a circuit, from which a netlist is created;

- MicroSim PSpice, the program that takes its output from MicroSim Schematics, describing the circuit topology, component values, and the desired analysis, and simulates the circuit; and

- MicroSim Probe, the program that reads a data file written by MicroSim PSpice describing the circuit behavior and permits you to plot the data.

Both MicroSim PSpice and MicroSim Probe were introduced in Chapter 1. There are several good references for MicroSim Schematics. MicroSim Schematics allows you to describe the circuit to be analyzed by drawing it, rather than using the MicroSim PSpice language illustrated in Chapter 1. But in order to exploit the power of MicroSim Schematics, you should be familiar with the MicroSim PSpice statements that are relevant to the simulation of linear electric circuits. Here we briefly outline the steps required to generate, simulate, and graphically examine a circuit:

1. Select each part from the libraries of available parts and place it on the screen.

2. As each part is placed, specify its attributes. Usually the only attribute that you must identify is the part's value.

3. Once all of the parts are arranged on the screen, connect the parts with wires.

---

[1]The examples that follow were generated using *The Student Edition of Schematics and MicroSim PSpice, Version 6.0.*

4. When the circuit is wired, add the circuit ground, which is a part from the parts libraries. The ground will be used as the reference node. Now your circuit is complete, and should be saved to a file.

5. Select the type of circuit analysis to be performed and describe the details of the analysis using the appropriate menus. Now your circuit is ready to simulate.

6. Simulation requires a netlist, which is produced automatically as the first stage of the simulation. If you have wired your circuit incorrectly, or have incorrectly specified attributes for any of the circuit components, an error message appears and the simulation halts. You must correct any errors before simulation can proceed.

7. The netlist serves as an input to MicroSim PSpice, which then attempts to simulate the behavior of the circuit according to the analysis parameters you have specified. If MicroSim PSpice encounters errors in the simulation, the simulation halts and you must correct the errors before you can proceed.

8. MicroSim PSpice produces a data file that is read by MicroSim Probe. MicroSim Probe then permits you to plot various circuit variables and explore the behavior of the circuit using a graphical interface.

The following examples suggest some of the applications suited for exploration using a schematic capture and circuit simulation tool.

EXAMPLE 2.1

Find the current through the resistor, as indicated in Fig. 2.1. This problem is taken from Problem 4.98 in the parent text.

SOLUTION

If you have attempted this problem by hand, you have discovered that it requires solving five simultaneous equations. Simulating the circuit is a good way to validate your hand solution, for it is easy to get confused solving five equations by hand.

The schematic is in the file \ds3\ex1.sch. When you load this file into MicroSim Schematics, you should see a screen similar to Fig. 2.2. The schematic in Fig. 2.2 is very similar to the circuit in Fig. 2.1, with two important differences. First, the schematic explicitly identifies the reference node. This is required for all schematics. While the circuit also contains a reference node, it is not explicitly indicated in Fig. 2.1. This is because the reference node can be identified as any node in the circuit convenient for circuit analysis. In your schematic,

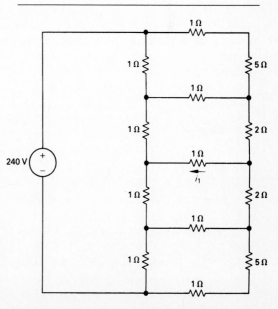

FIGURE 2.1 Circuit schematic for Example 2.1.

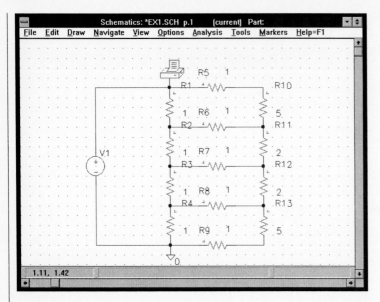

FIGURE 2.2 The circuit in Fig. 2.1, represented as a schematic.

you can define any node as the reference node, and all remaining voltages will be with respect to the reference node you select.

The other important difference between the schematic and the circuit is the little printer attached to the node at the top of the schematic. Click on the printer and examine its attributes, using the Attribute option of the Edit menu. You will see that the printer identifies which currents and voltages will be output once MicroSim PSpice analysis is complete. In this case, we see

```
analysis = dc  i(r_r7)
```

specifying that the current through the resistor named `r7` is to be printed to the output file. If you would like to see other resistor currents, or node voltages, you can specify them by editing the `analysis` attribute.

Select Setup from the Analysis menu to verify that DC Sweep has been selected. Click on DC Sweep to verify that a single dc voltage of 240 V has been specified. Then choose Simulate from the Analysis menu. This will create a netlist and automatically invoke MicroSim PSpice. Once MicroSim PSpice has completed the simulation, you can look at the output by selecting Examine Output from the MicroSim PSpice File menu. Page through the output until you come to the section titled DC Transfer Curves. You should see something similar to Fig. 2.3. Note that the current of interest is 0 A. If you specified other currents or voltages to be output, you should see their values as well.

To explore further, you might vary the value of the resistors, or allow the dc voltage source to vary over a range of values. Then you can print any current or voltage of interest versus the parameter you varied.

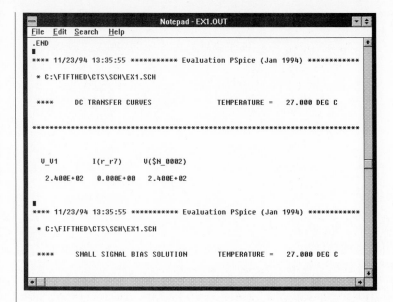

```
Notepad - EX1.OUT
File  Edit  Search  Help
.END

**** 11/23/94 13:35:55 ********** Evaluation PSpice (Jan 1994) ************

  * C:\FIFTHED\CTS\SCH\EX1.SCH


  ****     DC TRANSFER CURVES            TEMPERATURE =   27.000 DEG C



  ****************************************************************



    V_V1        I(r_r7)     V($N_0002)

    2.400E+02   0.000E+00   2.400E+02


**** 11/23/94 13:35:55 ********** Evaluation PSpice (Jan 1994) ************

  * C:\FIFTHED\CTS\SCH\EX1.SCH


  ****     SMALL SIGNAL BIAS SOLUTION    TEMPERATURE =   27.000 DEG C
```

FIGURE 2.3 Output from the simulation of the schematic in Fig. 2.2.

EXAMPLE 2.2

Suppose we wish to simulate a circuit containing an op amp. When using a circuit simulator, we have several choices. There are usually libraries containing models of actual electronic components such as op amps, and we can include one of these in our circuit. Often, we don't need a model that closely simulates the performance of an actual component. We may need only a simple equivalent circuit. In this case, we can construct this equivalent circuit and use it in place of the op amp in the circuit we wish to simulate.

Circuit simulators usually provide a third option, a method for creating your own circuit elements, to use in other circuits. To illustrate this in MicroSim PSpice, create a library model of an op amp, using the circuit in Fig. 2.4. This is the equivalent circuit discussed in Section 5.7 of the text. To make your op amp nearly ideal, use an input resistance $R_{in} = 1\,\text{M}\Omega$, an output resistance $R_o = 5\,\text{k}\Omega$, and a gain $A = 1 \times 10^6$.

SOLUTION

The MicroSim PSpice model of the equivalent circuit in Fig. 2.4 is in the file \ds3\nilsson.lib. A listing of this file is shown in Fig. 2.5. Later, you might want to examine the effect of using

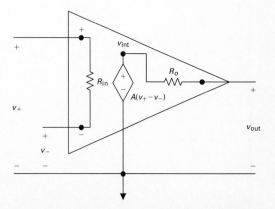

FIGURE 2.4 An equivalent circuit for an operational amplifier.

different values for the input and output resistances or for the gain. To do this, simply edit this file and make the desired changes.

```
.SUBCKT Ideal_OpAmp v+  v-  vout
*
*  This is a nearly ideal op amp
*
Rin        v+  v- 1MEG
Eamp       vint 0 v+ v- 1e6
Ro         vout vint 5k
.ENDS
```

FIGURE 2.5 Listing of the MicroSim PSpice file for the op amp model.

The schematic representation of this op amp model is in the file \ds3\nilsson.slb. You can examine this file in MicroSim Schematics by selecting Edit Library from the File menu, selecting Open from the next File menu, and selecting the file nilsson.slb. To display the symbol, select Get from the Part menu and select the part named Ideal_OpAmp. You should see this part, as in Fig. 2.6.

You tie this schematic to the model in the library by defining the part name to be the same as the part name in the library. Select Definition from the Part menu and note that the part name is Ideal_OpAmp, exactly as it appears in defining the SUBCKT shown in Fig. 2.5. Thus, any time you wish to create a new part, you must define a MicroSim PSpice model for that part, design a schematic symbol for the part, and tie the symbol and the model together using the definition of the symbol. In the next example, we use our newly created op amp part in a circuit.

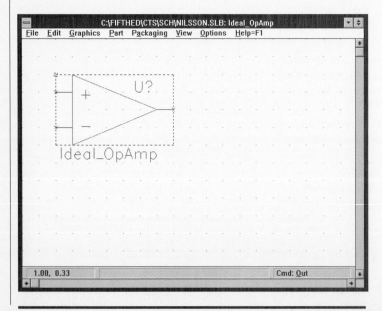

FIGURE 2.6 The schematic representation of the op amp model.

EXAMPLE 2.3

The circuit in Fig. 2.7 is an op-amp-based inverting amplifier, which is also shown in the parent text in Fig. 5.10.

a) Design an inverting amplifier with a gain of 5 for a load resistance of 10 kΩ.

b) Vary the resistance of the feedback resistor, R_f, from 1 kΩ to 9 kΩ in 2-kΩ steps and look at the amplification provided to the 10-kΩ load resistor.

c) Suppose the resistors are 5% resistors — how much can we expect the amplifier's gain to vary from the nominal design value of 5 in the worst case?

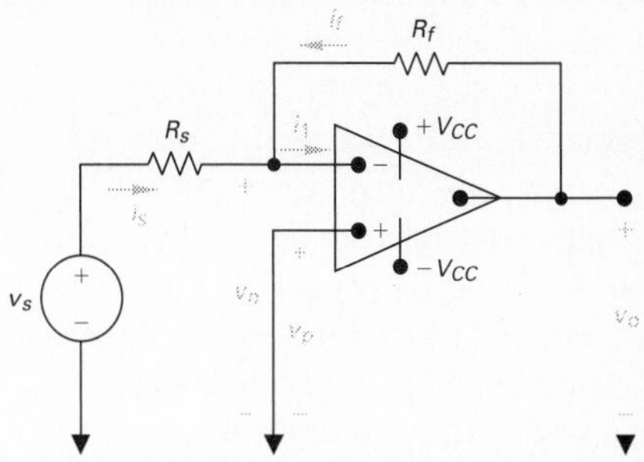

FIGURE 2.7 An op-amp-based inverting-amplifier circuit.

SOLUTION

a) A schematic for the circuit in Fig. 2.7 is in the file `\ds3 \ex3a.sch`. This schematic is shown in Fig. 2.8. We have added a load resistor (R_3) at the output of the op amp in order to measure the output voltage. Remember that the value of this load resistor has no effect on the gain of the amplifier. We have specified $R_3 = 10$ kΩ. Since the amplification is specified by the ratio of the feedback resistor to the input resistor (R_f/R_s), we have chosen a feedback resistor of 5 kΩ and an input resistor of 1 kΩ. Any two resistor values whose ratio is 5 will work — you can change the

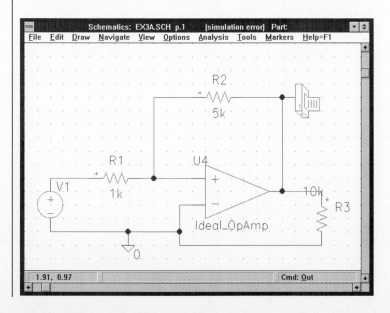

FIGURE 2.8 The schematic representation of the inverting amplifier shown in Fig. 2.7.

resistor values in the schematic at this point, so long as their ratio is maintained. Note that the schematic is nearly identical to the circuit in Fig. 2.7, but the schematic contains an explicit reference node, a load resistor, and the printer symbol that allows us to specify that the voltage at the output be printed.

If you look at the attributes of the voltage source, you will see that we have specified an ac source with a magnitude of 1 and a phase angle of 0 degrees. The analysis setup specifies an ac sweep, but the frequency is swept over only one value — we have chosen a frequency of 1 kHz, but you can choose any other frequency you like. Since the magnitude of the source is 1, the magnitude of the output will be the gain of the amplifier.

Part of the MicroSim PSpice output file is shown in Fig. 2.9. The magnitude of the voltage across the resistor R_3 is 5, as we have designed, while the phase angle is 180°. Remember that the ratio of the output voltage to the input voltage in an inverting amplifier is $-R_f/R_s$, so that the output is inverted, or shifted by 180° with respect to the input. Thus, our desired design objective was achieved with this choice of resistor values.

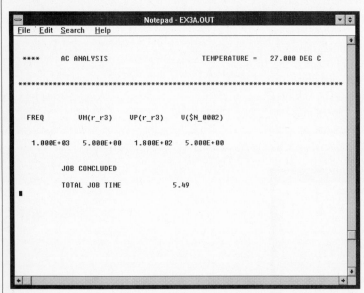

FIGURE 2.9 Output from the simulation of the schematic shown in Fig. 2.8.

b) To allow the feedback resistor values to be varied automatically by MicroSim PSpice and the simulation performed for each resistance value, we replace the feedback resistor with the resistor from the "breakout" library. The resistor's part name is RBREAK. The model details of components in the breakout library can be altered. We wish to alter the model of the feedback resistor to permit us to vary the resistance over the specified range.

You can delete the feedback resistor from the circuit shown in Fig. 2.8 and replace it with the part RBREAK. The resulting schematic is in the file \ds3\ex3b.sch and is shown in Fig. 2.10. We vary the value of the resistance in the RBREAK model by clicking on Parametric analysis in the Analysis Setup menu. The window that permits you to specify the values of the parameter is shown in Fig 2.11. If you would like to explore a different range of feedback resistor values, you can specify those at this point.

The circuit is simulated using the same analysis setup we used in part (a). Part of the output file produced is shown in Fig. 2.12. Here we see that if the resistance of RBREAK is 3 kΩ, the voltage

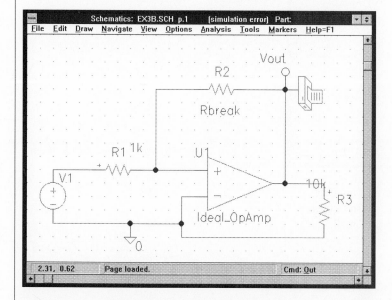

FIGURE 2.10 An inverting-amplifier schematic with a variable resistor in the feedback path.

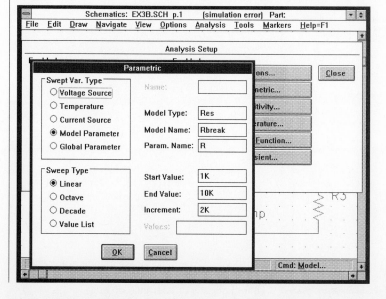

FIGURE 2.11 Specifying the parameters for the RBREAK resistor model.

output has a gain of 3 and a phase shift of $180°$, as you probably expected. The output contains values computed for resistances of 1, 3, 5, 7, and 9 kΩ, as you will see paging through it.

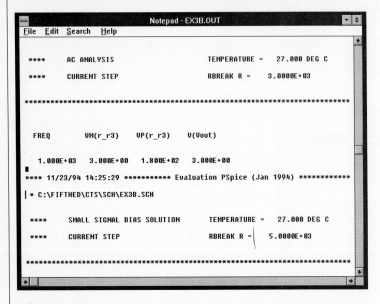

FIGURE 2.12 Output file from the simulation of the circuit in Fig. 2.10.

c) To examine the effect of resistor tolerance on our amplifier design, we replace both resistors with RBREAK resistors from the breakout library. You can modify the previous schematic, or use the schematic found in the file \ds3\ex3c.sch. The result is shown in Fig. 2.13.

To represent resistors with 5% tolerance, we change the definition of the RBREAK model. To do this, select Model from the Edit menu and edit the model definition. You will see that we have

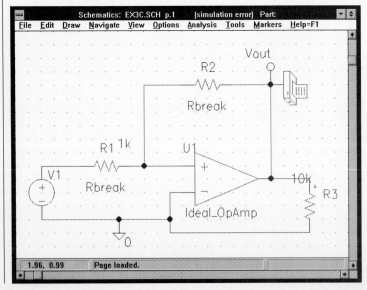

FIGURE 2.13 Schematic for the inverting-amplifier circuit of Fig. 2.7, prepared for examining the effect of resistor tolerance on gain.

redefined the RBREAK model to have a device tolerance of 5%, as shown in Fig. 2.14. You must define the RBREAK model this way for both resistors. Later, you may want to alter this definition to examine 1% resistors, or to look at the effect of a 5% or 1% lot tolerance instead of a device tolerance.

To simulate the circuit and examine the effect of resistor tolerance, we specify a worst-case analysis. We will perform two worst-case analyses — one to determine the maximum worst-case output voltage, and the other to determine the minimum worst-case output voltage. The parameters for the maximum worst-case analysis are shown in Fig. 2.15. Part of the output is shown in Fig. 2.16. You

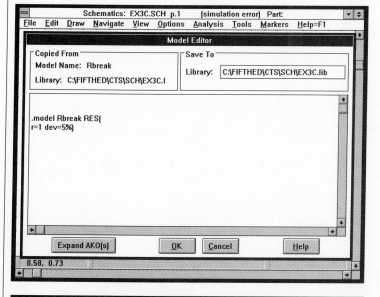

FIGURE 2.14 The new definition of the RBREAK model, used to examine the effect of resistors with 5% tolerance on gain.

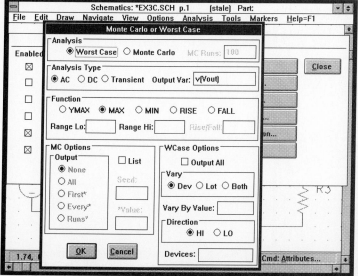

FIGURE 2.15 Parameter settings for the maximum worst-case analysis of the amplifier with 5% resistors.

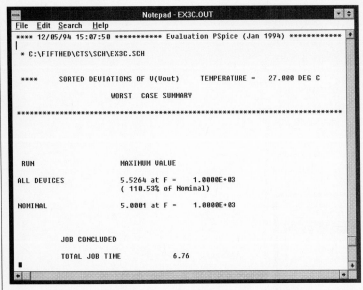

FIGURE 2.16 Output from the maximum worst-case analysis of the amplifier with 5% resistors.

can see that the maximum worst-case gain is 5.5264, or 110.53% greater than the gain of 5 specified in the design. If this were a real design problem, you would need to discuss the specification with the customer to determine whether or not such a variation from the design is acceptable.

The analysis parameters for the minimum worst-case gain are the same as those for the maximum worst-case gain, with the exception that the function MIN is selected and Vary by Value is LO, not HI. You must make these changes in MicroSim Schematics using the Setup option in the Analysis menu, and clicking on Monte Carlo/Worst Case. Once the changes are made, the screen should look like Fig. 2.17. Now simulate the analysis for this new

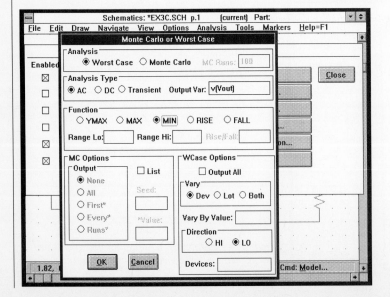

FIGURE 2.17 Parameter settings for the minimum worst-case analysis of the amplifier with 5% resistors.

parameter setting. Part of the resulting output file is shown in Fig. 2.18. The minimum worst-case gain is 4.5239, which is 90.476% of the designed gain of 5. Again, if this were a real circuit design, you would need to consult the customer to determine whether this variation is acceptable.

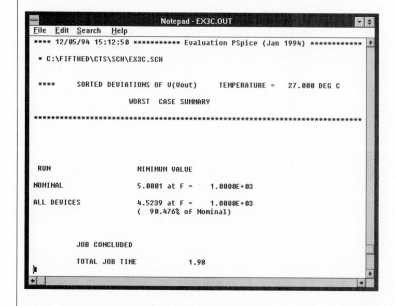

```
┌────────────────────────────────────────────────────────┐
│░░░           Notepad - EX3C.OUT              ░░▼░▲░     │
│ File  Edit  Search  Help                                │
│ **** 12/05/94 15:12:50 ********** Evaluation PSpice (Jan 1994) ************ │
│                                                         │
│  * C:\FIFTHED\CTS\SCH\EX3C.SCH                           │
│                                                         │
│  ****     SORTED DEVIATIONS OF V(Vout)    TEMPERATURE =  27.000 DEG C │
│                                                         │
│                  WORST  CASE SUMMARY                     │
│                                                         │
│  ***********************************************************************  │
│                                                         │
│                                                         │
│                                                         │
│   RUN                MINIMUM VALUE                       │
│                                                         │
│  NOMINAL              5.0001 at F =    1.0000E+03         │
│                                                         │
│  ALL DEVICES          4.5239 at F =    1.0000E+03         │
│                      (  90.476% of Nominal)              │
│                                                         │
│                                                         │
│          JOB CONCLUDED                                   │
│                                                         │
│          TOTAL JOB TIME          1.98                    │
│                                                         │
└────────────────────────────────────────────────────────┘
```

FIGURE 2.18 Output from the minimum worst-case analysis of the amplifier with 5% resistors.

E X A M P L E   2.4

The circuit shown in Fig. 2.19 is a cascade of two integrating amplifiers with feedback resistors. Suppose the component values are as follows: $R_a = 100$ kΩ, $R_1 = 500$ kΩ, $C_1 = 0.1$ μF, $R_b = 25$ kΩ, $R_2 = 100$ kΩ, and $C_2 = 1$ μF. The input voltage jumps from 0 to 250 mV at $t = 0$. At the instant the voltage is applied, no energy is stored in the capacitors. Plot the output voltage, $v_o(t)$. Note that this example is similar to Example 8.14 in the parent text.

S O L U T I O N

The circuit schematic is in the file \ds3\ex4.sch and is shown in Fig. 2.20. It employs the ideal op amp model constructed in Example 2.2, and uses a dc source to model the supplied voltage. To plot the

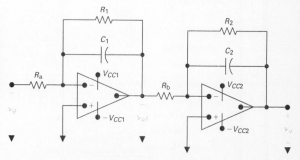

FIGURE 2.19 Cascade of integrating amplifiers with feedback resistors.

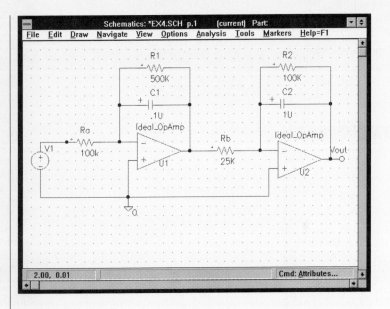

FIGURE 2.20 Schematic of the circuit in Fig. 2.19.

output voltage as a function of time, we specify Transient Analysis
in the Analysis setup. We need to calculate the length of time for the
analysis, which depends on the time constants of the circuit. There are
two time constants — $\tau_1 = R_1 C_1 = 0.05$ s, and $\tau_2 = R_2 C_2 = 0.10$ s.
We choose a final time of 0.5 seconds, five times larger than the larger
time constant. The parameters for the transient analysis are shown in
Fig. 2.21. If you are interested in continuing the analysis beyond 0.5 s,
you can modify the Final Time value.

After MicroSim PSpice finishes the circuit simulation, Microsim

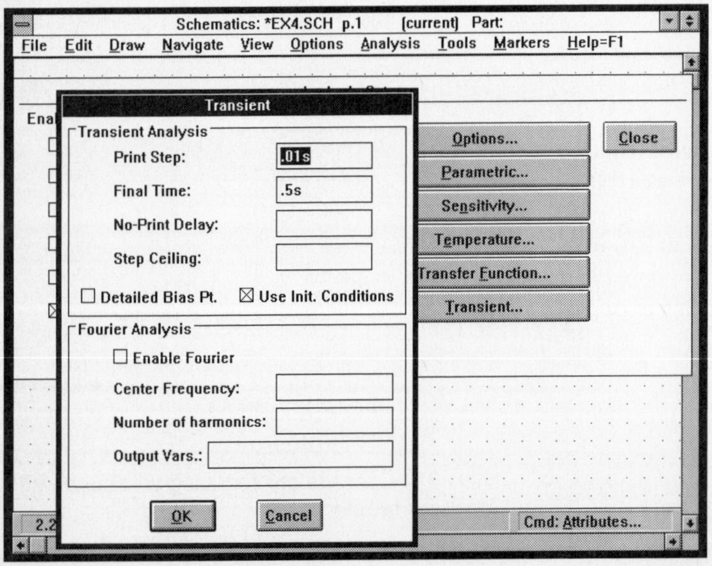

FIGURE 2.21 Parameter settings for the transient
analysis of the circuit schematic in Fig. 2.20.

Probe is automatically invoked to provide a graphical interface to the results. The plot of the output voltage is shown in Fig. 2.22.

One important use of circuit simulation is validating your analytical results. In Example 8.14 in the parent text, analysis of the circuit in Fig. 2.19 yields the following closed-form expression for the output voltage:

$$v_o(t) = 5 - 10e^{-10t} + 5e^{-20t}.$$

To validate this result, we plot this equation on the MicroSim Probe graph of $v_o(t)$. You must use the MicroSim Probe functions `exp` and `time`, as shown in the label at the bottom of the graph. As you can see from Fig. 2.22, the two plots lie directly on top of one another, validating the analytical expression for $v_o(t)$.

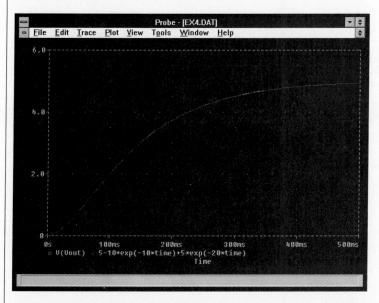

FIGURE 2.22 Output voltage from the transient analysis of the circuit schematic in Fig. 2.20.

EXAMPLE 2.5

The series *RLC* bandpass filter is analyzed in Section 15.4 of the text. In order to gain a better understanding of the effect of the capacitor value on the bandpass behavior, plot the magnitude characteristics for a series *RLC* bandpass filter with $R = 8$ kΩ, $L = 40$ mH, and C varying from 0.15 μF to 0.35 μF in 0.05 μF increments.

SOLUTION

The circuit schematic is found in the file \ds3\ex5.sch and is shown in Fig. 2.23. The values of resistance and inductance are fixed, but since we will vary the value of capacitance, we use the capacitor

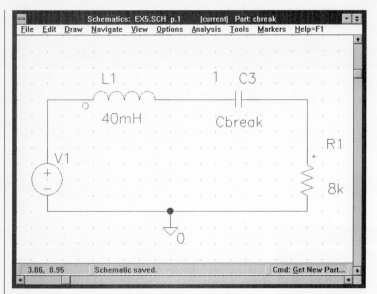

FIGURE 2.23 Schematic for the series *RLC* bandpass filter with variable capacitance.

from the breakout library, called CBREAK. Two types of analyses must be specified in the analysis setup. One is parametric, which will vary the capacitor value throughout the desired range, simulating the circuit for each value. Parameters for this analysis are shown in Fig. 2.24.

We must also specify ac sweep analysis, in order to assess the frequency response properties of this circuit. To produce a plot in the familiar Bode form, you must specify the frequency in decades. We need to ensure that the decades chosen include those where the circuit's behavior is changing. For a bandpass filter, this means including the decade that contains the center frequency, as well as several decades on either side of the center frequency. The center frequency of the

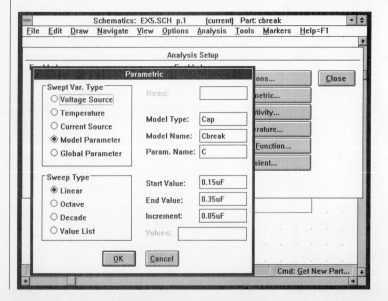

FIGURE 2.24 Parametric variation of the capacitor value for the series *RLC* bandpass filter.

series RLC circuit is $\sqrt{1/LC}$. Since both MicroSim PSpice and MicroSim Probe deal with frequency in Hz, the center frequency will vary from 1345.1 Hz (for $C = 0.15$ μF) to 2054.68 Hz (for $C = 0.35$ μF). The bandwidth is given by R/L, which is about 31.8 kHz. As you can see from Fig. 2.25, we choose a range of frequencies from 10 Hz to 1 MHz.

Once MicroSim PSpice has simulated the circuit for the values of capacitance, MicroSim Probe is automatically invoked so that you may examine the results of the simulation graphically. Although we could plot the results for any one of the capacitor values separately, we choose to plot the results for all five capacitor values. The resulting plot is shown in Fig. 2.26. At this point you might wish to investigate

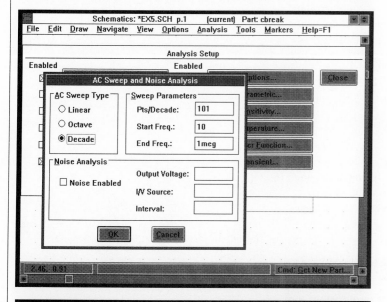

FIGURE 2.25 Parameters for the ac sweep of the series RLC bandpass filter.

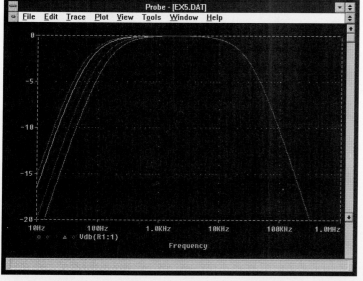

FIGURE 2.26 Frequency response plot for the RLC bandpass filter.

the frequency response behavior of some other circuit variable, such as the current through the circuit or the voltage drop across the inductor. This is easy to do with MicroSim Probe.

EXAMPLE 2.6

Suppose that the pulsed waveform shown in Fig. 2.27(a) is input to the *RC* circuit shown in Fig. 2.27(b). Simulate the circuit for two values of capacitance, $C = 1 \, \mu\text{F}$ and $C = 10 \, \mu\text{F}$, and plot the input voltage and the resulting output voltage for each capacitor value. Examine the frequency content of the input and output voltages and comment. This example provides a specific analysis of the circuit studied in Section 17.5 in the parent text.

SOLUTION

The schematic for the *RC* circuit is in the file `\ds3\ex6.sch` and is shown in Fig. 2.28. We use a fixed resistor and the capacitor from the breakout library. We specify the voltage source as a pulsed source, and set its attributes using the Attribute option from the Edit menu. The attribute settings are given in Fig. 2.29. If you wish to examine a different pulsed source, you can alter the settings from this menu.

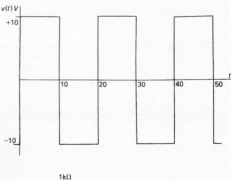

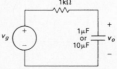

FIGURE 2.27 (a) A pulsed-voltage waveform; (b) an *RC* circuit with the pulsed-voltage waveform input.

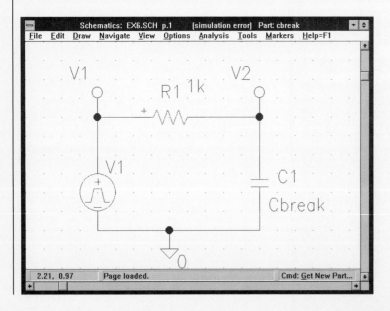

FIGURE 2.28 Schematic for the *RC* circuit from Fig. 2.27(b).

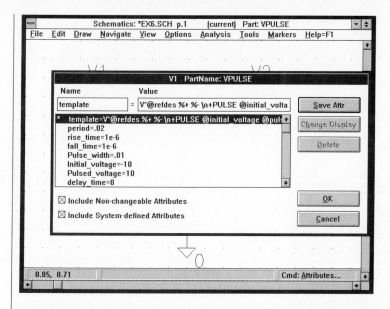

FIGURE 2.29 Attribute settings for the pulsed-voltage source shown in Fig. 2.27(a).

You must specify two types of analysis. Parametric analysis will allow the capacitor value to vary, and MicroSim PSpice will simulate the resulting circuit for each value. Transient analysis will enable us to examine the effect of the RC circuit on the pulsed source as we look at the output voltage across the capacitor. The Transient selection will also permit us to specify that the coefficients of the Fourier series components be written to the output file. Parameters specifying transient analysis and Fourier analysis are shown in Fig. 2.30.

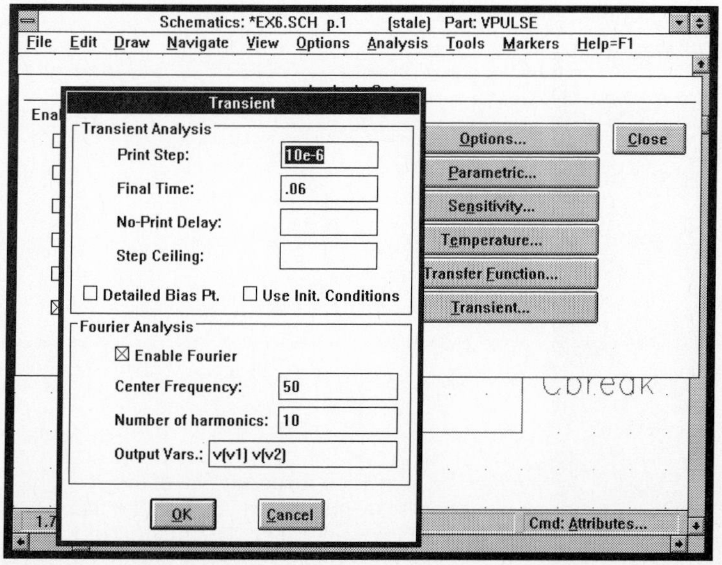

FIGURE 2.30 Transient analysis parameters for the *RC* circuit with pulsed voltage source.

Following MicroSim PSpice simulation of the schematic in Fig. 2.28, MicroSim Probe is automatically invoked. We choose to examine the circuit with a 1 μF capacitor first. A plot of both the input and output voltage versus time shows that the output is able to follow the input, due to the relatively small value of the capacitance. We can examine the Fourier components of both the input and output waveforms by selecting the Plot menu and choosing Fourier from the X-Axis option. Both the time response plot and the Fourier component plot are shown in Fig. 2.31. The Fourier component plot verifies that the

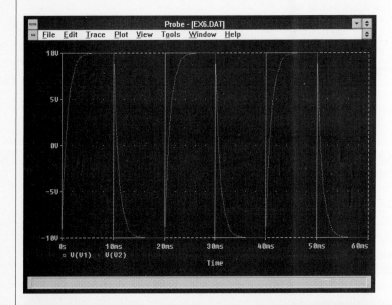

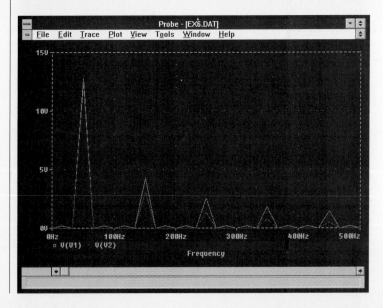

FIGURE 2.31 (a) Time response plot of the pulsed-voltage input of Fig. 2.27(a) and the output of the circuit in Fig. 2.27(b) for $C = 1\ \mu$F; (b) Fourier components for the waveforms shown in part (a).

frequency content of the output waveform is nearly the same as the frequency content of the input waveform.

The Fourier coefficients of the input waveform are in the file output from MicroSim PSpice simulation. Part of this file is shown in Fig. 2.32. Compare these component values to those for the output waveform, also found in the file output from MicroSim PSpice simulation, and shown in Fig. 2.33. The magnitudes at each frequency are very similar.

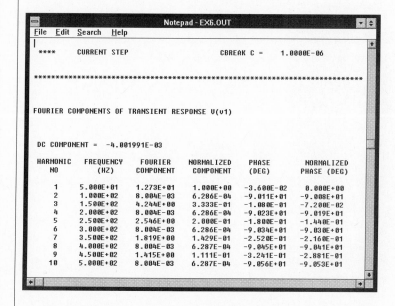

```
                        Notepad - EX6.OUT
 File  Edit  Search  Help
|
 ****      CURRENT STEP              CBREAK C =    1.0000E-06


 **************************************************************************


 FOURIER COMPONENTS OF TRANSIENT RESPONSE U(v1)


 DC COMPONENT =  -4.001991E-03

 HARMONIC   FREQUENCY    FOURIER    NORMALIZED    PHASE      NORMALIZED
   NO         (HZ)      COMPONENT   COMPONENT     (DEG)     PHASE (DEG)

    1       5.000E+01   1.273E+01   1.000E+00   -3.600E-02   0.000E+00
    2       1.000E+02   8.004E-03   6.286E-04   -9.011E+01  -9.008E+01
    3       1.500E+02   4.244E+00   3.333E-01   -1.080E-01  -7.200E-02
    4       2.000E+02   8.004E-03   6.286E-04   -9.023E+01  -9.019E+01
    5       2.500E+02   2.546E+00   2.000E-01   -1.800E-01  -1.440E-01
    6       3.000E+02   8.004E-03   6.286E-04   -9.034E+01  -9.030E+01
    7       3.500E+02   1.819E+00   1.429E-01   -2.520E-01  -2.160E-01
    8       4.000E+02   8.004E-03   6.287E-04   -9.045E+01  -9.041E+01
    9       4.500E+02   1.415E+00   1.111E-01   -3.241E-01  -2.881E-01
   10       5.000E+02   8.004E-03   6.287E-04   -9.056E+01  -9.053E+01
```

FIGURE 2.32 Fourier components of the pulsed-voltage waveform in Fig. 2.27(a).

```
                        Notepad - EX6.OUT
 File  Edit  Search  Help
 ****      CURRENT STEP              CBREAK C =    1.0000E-06


 **************************************************************************


 FOURIER COMPONENTS OF TRANSIENT RESPONSE U(v2)


 DC COMPONENT =   9.914499E-04

 HARMONIC   FREQUENCY    FOURIER    NORMALIZED    PHASE      NORMALIZED
   NO         (HZ)      COMPONENT   COMPONENT     (DEG)     PHASE (DEG)

    1       5.000E+01   1.217E+01   1.000E+00   -1.755E+01   0.000E+00
    2       1.000E+02   1.685E-03   1.384E-04    5.694E+01   7.449E+01
    3       1.500E+02   3.083E+00   2.533E-01   -4.425E+01  -2.670E+01
    4       2.000E+02   1.223E-03   1.004E-04    3.663E+01   5.418E+01
    5       2.500E+02   1.348E+00   1.107E-01   -5.838E+01  -4.083E+01
    6       3.000E+02   9.147E-04   7.513E-05    2.590E+01   4.345E+01
    7       3.500E+02   7.458E-01   6.126E-02   -6.656E+01  -4.901E+01
    8       4.000E+02   7.195E-04   5.910E-05    1.847E+01   3.602E+01
    9       4.500E+02   4.581E-01   3.763E-02   -7.310E+01  -5.555E+01
   10       5.000E+02   5.634E-04   4.628E-05    1.333E+01   3.088E+01
```

FIGURE 2.33 Fourier components of the output voltage waveform for the *RC* circuit, with $C = 1 \ \mu F$.

Now examine the response of the circuit when $C = 10 \ \mu$F. A plot of both the input and output voltage versus time shows that the larger value of capacitance makes it difficult for the output to follow the pulsed signal. When we examine the Fourier components of both the input and output waveforms, we find that they are quite different. Both the time response plot and the Fourier component plot are shown in Fig. 2.34.

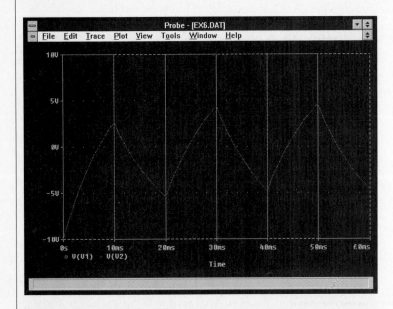

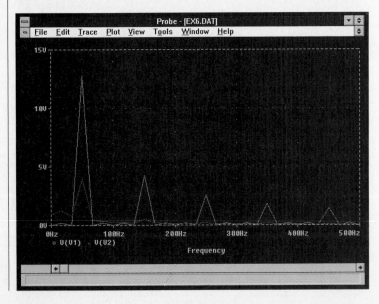

FIGURE 2.34 (a) Time response plot of the pulsed-voltage input of Fig. 2.27(a) and the output of the circuit in Fig. 2.27(b) for $C = 10 \ \mu$F; (b) Fourier components for the waveforms shown in part (a).

The Fourier coefficients of the input waveform are the same as before, since changing the capacitor value has no effect on the pulsed-voltage input. Compare these component values (Fig. 2.32) to those for the output waveform, shown in Fig. 2.35. As you probably suspected, the component values for the input and output are quite different.

This simulation confirms the analysis in Section 17.5 in the parent text, and adds insight due to the graphical representation of the circuit's behavior. This is another important application of circuit simulation software to the study of electric circuits.

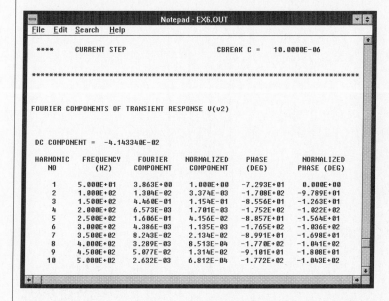

FIGURE 2.35 Fourier components of the output voltage waveform for the *RC* circuit, with $C = 10 \mu F$.

BIBLIOGRAPHY

Marc E. Herniter, *Schematic Capture with MicroSim PSpice*, Merrill/Macmillan Publishing Co., NY, 1994.

John Keown, *MicroSim PSpice and Circuit Analysis*, second ed., Merrill/Macmillan Publishing Co., NY, 1993.

MicroSim Corporation, *The Design Center: Schematic Capture — User's Guide*, Irvine, CA, 1994.

3 MATRIX EQUATION SOLVERS

One of the most useful mathematical forms for modeling a system with multiple describing equations is a matrix. In circuit analysis, we use matrix equations to describe circuits with multiple node voltage equations or multiple mesh current equations. Then we can use tools such as Cramer's method to solve for the unknowns in the matrix equation. But as you have probably discovered, the methods used to solve matrix equations quickly become unwieldy when the matrix is larger than three-by-three.

Matrix equations are found in all branches of engineering, science, and mathematics. It is not surprising, therefore, that several computational tools have been developed that automate the solution of matrix equations. These computational tools require that the model of the system under study, in our case a linear circuit, be described using a matrix equation. The computer can then quickly solve the equation, even if it has more than three dimensions. This computational support enables us to study and analyze circuits that are described by four or more equations and would be tedious to work with by hand. Further, the computer enables us to consider changing various parameters in the circuit, solving the circuit again, and learning how these changes impact the circuit's behavior. Finally, many of the matrix analysis tools have graphical interfaces, which allow us to visualize the solution and perhaps gain a deeper understanding of the circuit.

The particular software tool we use as an example here is MATLAB®, developed by The MathWorks, Inc.[1] The name is a shortened form of "matrix laboratory," which is indeed an appropriate name for this popular package. MATLAB can be useful in a wide variety of applications and is a very powerful tool. Here we focus on a

---

[1]The examples that follow were generated using *The Student Edition of MATLAB for Windows, Version 4.0.*

subset of MATLAB features that have particular application in linear circuit analysis. Thus, there are many important MATLAB functions that we will not examine.

You can use MATLAB in two different ways: (1) in the interactive mode, and (2) by running an M-file. If you are solving a problem on a one-time basis, and you don't intend to solve this problem or a more general problem ever again, you can use MATLAB to generate the solution interactively. Once you see the MATLAB prompt, you can specify the matrix and any other information required, and MATLAB will compute and display the solution for you. If you wish to solve the problem again, you will need to specify all of the information again. You might wish to use MATLAB in this interactive mode to confirm a circuit analysis result you have developed by hand.

In this chapter, we specify problems for MATLAB using M-files. An M-file is simply a list of MATLAB commands that define a problem for MATLAB and request a solution. You can create an M-file using a text editor of your choice, and it is so named because the filename containing the MATLAB commands must have the extension `.m`. When you want to execute the commands in the M-file, you simply type the name of the M-file at the MATLAB prompt. If you want to re-run the commands, solving the problem a second time, you just type the name of the M-file at the MATLAB prompt again. You can make changes to the M-file in the editor and re-run the altered file. Perhaps the most powerful use of the M-file is to have it prompt the user for data, so that the problem in the M-file solves a general, rather than a specific, problem. We will illustrate several general-problem M-files.

The examples below illustrate some of the functions provided in MATLAB that can be used in circuit analysis. Because MATLAB is not specifically designed to support circuit analysis, a good deal of preliminary work is usually needed to transform a circuit analysis problem into a form suitable for MATLAB. This generally means that you have to construct the circuit's describing equations and formulate the problem as a matrix equation before MATLAB can begin a solution. We illustrate this transformation from circuit problem to MATLAB problem in the examples.

There are several good introductory references that introduce the MATLAB functions. It is assumed in what follows that you are familiar with elementary MATLAB functions. As a short reminder, the general steps for solving a circuit problem using MATLAB are as follows:

1. Write down the circuit equations that include the variables of interest in the circuit problem.

2. Express the circuit equations in matrix form:

$$x = Ay,$$

where y is an n-dimensional vector of independent variables, x is an m-dimensional vector of dependent variables, and A is an n-by-m matrix that relates the independent and dependent variables.

3. Use an editor to construct an M-file that solves the matrix. The M-file usually consists of three parts:

   * The input, which prompts the user to input specific values for the matrix equation;

   * The solution, which specifies the MATLAB tools used to solve the matrix equation; and

   * The result, which displays the numerical or graphical result of the solution.

4. Execute the M-file by typing its name at the MATLAB prompt.

EXAMPLE 3.1

Find the current i_1 in the circuit shown in Fig. 3.1. This is Problem 4.98 in the parent text.

SOLUTION

We begin by defining the mesh currents for this circuit, as shown in Fig. 3.2. Now we can write the five mesh equations for this circuit:

$$240 = 4i_a - i_b - i_c - i_d - i_e;$$
$$0 = -i_a + 8i_b - i_c;$$
$$0 = -i_a - i_b + 5i_c - i_d;$$
$$0 = -i_a - i_c + 5i_d - i_e;$$
$$0 = -i_a - i_d + 8i_e.$$

Since MATLAB requires these equations to be specified in the form of a matrix, we write

$$\begin{bmatrix} 240 \\ 0 \\ 0 \\ 0 \\ 0 \end{bmatrix} = \begin{bmatrix} 4 & -1 & -1 & -1 & -1 \\ -1 & 8 & -1 & 0 & 0 \\ -1 & -1 & 5 & -1 & 0 \\ -1 & 0 & -1 & 5 & -1 \\ -1 & 0 & 0 & -1 & 8 \end{bmatrix} \begin{bmatrix} i_a \\ i_b \\ i_c \\ i_d \\ i_e \end{bmatrix}.$$

MATLAB can solve this matrix equation interactively. Try this by typing in the five-by-five matrix and the numerical vector, and asking MATLAB to calculate the ratio of the matrix to the vector. You will see on the display the calculated vector of currents, from which you can calculate the current i_1 specified in Fig. 3.1.

We created an M-file that specifies the matrix equation, the solution method, and the result displayed. You can find this M-file in

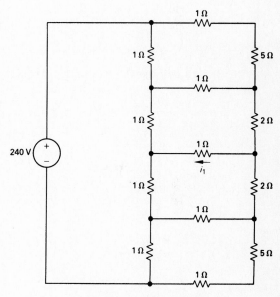

FIGURE 3.1 The circuit for Example 3.1.

the \matlab subdirectory with the filename ex1.m. It is listed in
Fig. 3.3.

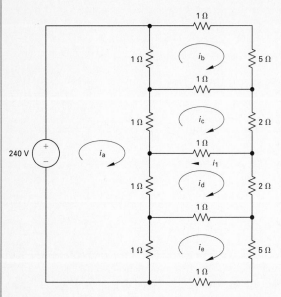

FIGURE 3.2 **The circuit in Fig. 3.1 with mesh currents assigned.**

```
% This example solves for the currents in the
% circuit shown in Fig. 3.1
fprintf('The mesh current matrix is \n')
R = [ 4, -1, -1, -1, -1;
     -1,  8, -1,  0,  0;
     -1, -1,  5, -1,  0;
     -1,  0, -1,  5, -1;
     -1,  0,  0, -1,  8]
Vin = [ 240, 0, 0, 0, 0]';
fprintf('The voltage vector is\n')
V = Vin'
Iout = R\Vin;
fprintf('The mesh current vector is \n')
I = Iout'
I1 = Iout(3)-Iout(4);
fprintf('The current I1 is %3.2f \n', I1)
```

FIGURE 3.3 Listing of the M-file for the circuit in Fig. 3.1.

In this M-file, we define and display the mesh-current matrix and
the voltage vector, compute and display the mesh-current vector, and
compute and display the current i_1, which is defined as the difference
between the current i_c and the current i_d, as shown in Fig. 3.2. A
sample of the screen output when this M-file is executed is shown
in Fig. 3.4. Note that we printed the current and voltage vectors as

row vectors, rather than column vectors, so that they would both fit on the screen. We see that the computed current i_1 is 0 A.

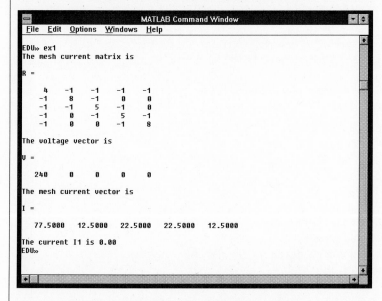

```
                                    MATLAB Command Window
 File   Edit   Options   Windows   Help

EDU» ex1
The mesh current matrix is

R =

    4     -1    -1    -1    -1
   -1      8    -1     0     0
   -1     -1     5    -1     0
   -1      0    -1     5    -1
   -1      0     0    -1     8

The voltage vector is

V =

  240      0     0     0     0

The mesh current vector is

I =

   77.5000   12.5000   22.5000   22.5000   12.5000

The current I1 is 0.00
EDU»
```

FIGURE 3.4 The result of executing the M-file listed in Fig. 3.3.

EXAMPLE 3.2

Calculate the current i_1 shown in Fig. 3.5 for the resistor values and voltage source value used in Example 3.1 (Fig. 3.1). Then recalculate the current i_1 for the following resistor and voltage source values: $R_1 = R_2 = R_3 = R_4 = R_5 = R_6 = R_7 = R_8 = R_9 = 2\Omega$; $R_{10} = R_{13} = 20\ \Omega$; $R_{11} = R_{12} = 5\ \Omega$; $V_1 = 100$ V. This is an extension of Problem 4.98 in the text.

SOLUTION

Even though this circuit is the same as the one in Example 3.1, we take a slightly different approach. Since we need to solve the circuit multiple times, for different resistor values and voltage source values, we will generate the matrix equation that describes the circuit for any values of resistors and voltage source. We begin by assigning the mesh currents, as we did in Example 3.1 and show in Fig. 3.6. From this circuit we generate the following mesh equations:

$$V_1 = (R_1 + R_2 + R_3 + R_4)i_a - R_1i_b - R_2i_c - R_3i_d - R_4i_e;$$
$$0 = -R_1i_a + (R_1 + R_5 + R_6 + R_{10})i_b - R_6i_c;$$
$$0 = -R_2i_a - R_6i_b + (R_2 + R_6 + R_7 + R_{11})i_c - R_7i_d;$$
$$0 = -R_3i_a - R_7i_c + (R_3 + R_7 + R_8 + R_{12})i_d - R_8i_e;$$
$$0 = -R_4i_a - R_8i_d + (R_4 + R_8 + R_9 + R_{13})i_e.$$

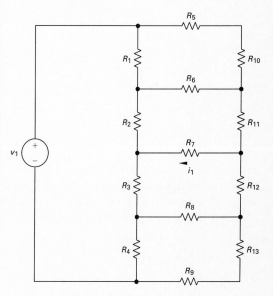

FIGURE 3.5 The circuit for Example 3.2.

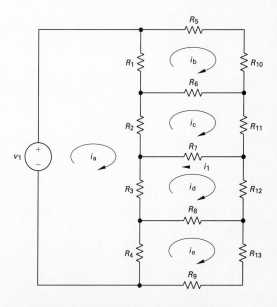

FIGURE 3.6 The circuit in Fig. 3.5 with mesh currents assigned.

We now write these equations in matrix form, as required by MATLAB:

$$
\begin{bmatrix} V_1 \\ 0 \\ 0 \\ 0 \\ 0 \end{bmatrix} =
\begin{bmatrix}
R_1 + R_2 \\ +R_3 + R_4 & -R_1 & -R_2 & -R_3 & -R_4 \\
-R_1 & \begin{array}{c}R_1 + R_5 \\ +R_6 + R_{10}\end{array} & -R_6 & 0 & 0 \\
-R_2 & -R_6 & \begin{array}{c}R_2 + R_6 \\ +R_7 + R_{11}\end{array} & -R_7 & 0 \\
-R_3 & 0 & -R_7 & \begin{array}{c}R_3 + R_7 \\ +R_8 + R_{12}\end{array} & -R_8 \\
-R_4 & 0 & 0 & -R_8 & \begin{array}{c}R_4 + R_8 \\ +R_9 + R_{13}\end{array}
\end{bmatrix}
\begin{bmatrix} i_a \\ i_b \\ i_c \\ i_d \\ i_e \end{bmatrix}.
$$

Now we create an M-file that specifies this matrix equation, allows the user to input the values for the resistors and the voltage source, computes the mesh-current vector for these values, and outputs the computed current i_1, which is the difference between currents i_c and i_d. The M-file in the file \matlab\ex2.m, and is listed in Fig. 3.7.

```
%
%  This program reads component values for the circuit in
%  Example 3.2, then computes the mesh currents
%
R = input('Input resistor values: [R1, R2, R3, R4, R5, R6,
R7, R8, R9, R10, R11, R12, R13] ');
V = input('Input independent voltage, [V] ');
%
A = [ R(1)+R(2)+R(3)+R(4) , -R(1) , -R(2) , -R(3) , -R(4) ;
     -R(1) , R(1)+R(5)+R(6)+R(10) , -R(6) , 0 , 0 ;
     -R(2) , -R(6) , R(2)+R(6)+R(7)+R(11) , -R(7) , 0 ;
     -R(3) , 0 , -R(7) , R(3)+R(7)+R(8)+R(12) , -R(8) ;
     -R(4) , 0 , 0 , -R(8) , R(4)+R(8)+R(9)+R(13) ];
B = [ V(1) , 0, 0, 0, 0 ]';
%
fprintf('The mesh currents are \n')
i = A\B
fprintf('The current I1 is %3.2f \n', i(3)-i(4))
```

FIGURE 3.7 A listing of the M-file that solves the matrix equation describing the circuit in Fig. 3.5.

This M-file now solves the general circuit shown in Fig. 3.5 for any values of resistance and voltage source. Figure 3.8 shows the result of solving the circuit for the resistor and voltage source values used

```
┌──────────────────────────────────────────────────────────┐
│ ▦                    MATLAB Command Window         ▼ ◆     │
│  File   Edit   Options   Windows   Help                    │
│ EDU» ex2                                                   │
│                                                            │
│ Input R values: [R1, ..., R13] [1 1 1 1 1 1 1 1 1 5 2 2 5] │
│                                                            │
│ Input independent voltage, [V] 240                         │
│ The mesh currents are                                      │
│                                                            │
│ i =                                                        │
│                                                            │
│     77.5000                                                │
│     12.5000                                                │
│     22.5000                                                │
│     22.5000                                                │
│     12.5000                                                │
│                                                            │
│ The current I1 is 0.00                                     │
│ EDU»                                                       │
│                                                            │
└──────────────────────────────────────────────────────────┘
```

FIGURE 3.8 Output from the M-file in Fig. 3.7 for the component values in Example 3.1.

in Example 3.1. Figure 3.9 shows the result when the component values specified in this example are input. Note that in both cases, the current i_1 is zero. Try using this M-file to determine whether i_1 is zero for all component values, or whether the component values must be related to one another in a certain way to guarantee that $i_1 = 0$ A.

```
┌──────────────────────────────────────────────────────────┐
│ ▦                    MATLAB Command Window         ▼ ◆     │
│  File   Edit   Options   Windows   Help                    │
│ EDU» ex2                                                   │
│                                                            │
│ Input R values: [R1, ..., R13] [2 2 2 2 2 2 2 2 2 20 5 5 20]│
│                                                            │
│ Input independent voltage, [V] 100                         │
│ The mesh currents are                                      │
│                                                            │
│ i =                                                        │
│                                                            │
│     15.0524                                                │
│      1.4398                                                │
│      3.6649                                                │
│      3.6649                                                │
│      1.4398                                                │
│                                                            │
│ The current I1 is 0.00                                     │
│ EDU» |                                                     │
│                                                            │
└──────────────────────────────────────────────────────────┘
```

FIGURE 3.9 Output from the M-file in Fig. 3.7 for a different set of component values.

E X A M P L E   3.3

The current pulse applied to the 100-mH inductor shown in Fig. 3.10 is 0 for $t < 0$ and is given by the expression

$$i(t) = 10te^{-5t}$$

for $t > 0$. Plot the inductor voltage and current as a function of time. This circuit was explored analytically in Example 6.1 in the text.

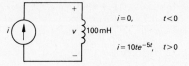

FIGURE 3.10 The circuit for Example 3.3.

S O L U T I O N

MATLAB has many different mathematical tools that are useful in circuit analysis. In this problem, we will use the exp() function in specifying the current pulse, and will use the diff() function to compute the derivative of the current as the ratio of the forward difference of the voltage current to the forward difference of the time vector, or di/dt.

A listing of the M-file that specifies the current pulse, computes the inductor voltage, and plots both is shown in Fig. 3.11.

```
% Plot voltage and current across an inductor
t = 0:0.01:1;
i = 10*t.*exp(-5*t);
L = input('Enter inductance L in henrys: ');
di = L*diff(i)./diff(t);
td = t(2:length(t));
subplot(2,1,1), plot(t, i'w')
grid
xlabel('t (sec)')
ylabel('i (A)')
subplot(2,1,2), plot(td, di'w')
grid
xlabel('t (sec)')
ylabel('v (V)')
```

FIGURE 3.11 A listing of the M-file that specifies the solution for the problem in Example 3.3.

This file is in the \matlab subdirectory and has the name ex3.m. Note that we have chosen to allow the user to input the inductance, so that you can explore the effect of inductor value on the voltage and current plots. An example of the output plots for a 100-mH inductor is shown in Fig. 3.12.

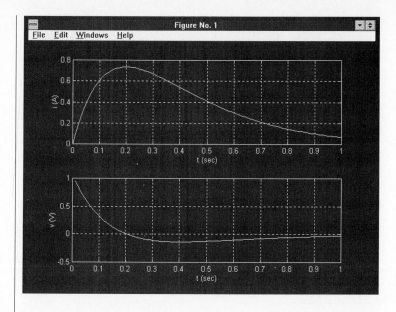

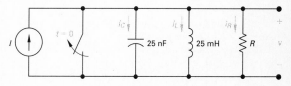

FIGURE 3.12 Output plots from the M-file in Fig. 3.11 for a 100-mH inductor.

Feel free to modify this M-file to examine the voltage and current plots for different forms of current pulses. MATLAB's ability to generate plots quickly for many different current pulses may enable you to gain a greater appreciation for the behavior of inductors in the face of different forcing functions. You might also change the current source in series with an inductor to a voltage source in parallel with a capacitor and use MATLAB to explore this simple but important circuit.

EXAMPLE 3.4

Use MATLAB to plot the response of the inductor current in the parallel *RLC* circuit in Fig. 3.13. Assume that the source current is a step of 100 mA, the value of the inductor is 25 mH, and the value of the capacitor is 25 nF. Explore the effect of the resistor value on the inductor current of this circuit by choosing three different values of resistance — 400 Ω, 500 Ω, and 625 Ω. Plot your results. (Note that this example is essentially the same as Example 8.9 of the parent text.)

FIGURE 3.13 The circuit for Example 3.4.

SOLUTION

We begin by deriving the describing equation for the circuit in Fig. 3.13. The describing equation is a differential equation, and its derivation is presented in Chapter 8 of the parent text. The result is the second-order differential equation

$$\frac{d^2 i_L}{dt^2} + \frac{1}{RC}\frac{di_L}{dt} + \frac{i_L}{LC} = \frac{I}{LC}.$$

MATLAB cannot solve second-order differential equations directly. Instead, you must represent the second-order (or any higher-order) equation as a set of first-order equations. This is easy to do using a change of variables. Let $i_1 = i_L$ and $i_2 = di_L/dt$. Then we can write the second-order differential equation as a set of two first-order differential equations, as follows:

$$\frac{di_1}{dt} = \frac{di_L}{dt} = i_2;$$

$$\frac{di_2}{dt} = \frac{d^2 i_L}{dt^2} = \frac{I}{LC} - \frac{i_2}{RC} - \frac{i_1}{LC}.$$

Once the higher-order differential equation is transformed into a set of first-order differential equations, we can use the MATLAB function ode23() to solve the set of equations. The M-file containing this MATLAB function is in the \matlab subdirectory and has the filename ex4.m. This file is listed in Fig. 3.14. Note that the user is

```
% look at the inductor current in a parallel RLC circuit
global R_times_C  L_times_C  I_source
R = input('Input the resistance value in ohms, ');
L = input('Input the inductance value in henrys, ');
C = input('Input the capacitance value in farads, ');
I_source = input('Input the value of the current source, in amps, ');
R_times_C = R*C;
L_times_C = L*C;
initial = [0 0];
[t1,num1_i] = ode23('ex4eq2', 0, 0.0002, initial);
R = input('Input a second resistance value in ohms, ');
R_times_C = R*C;
[t2,num2_i] = ode23('ex4eq2', 0, 0.0002, initial);
R = input('Input a third resistance value in ohms, ');
R_times_C = R*C;
[t3,num3_i] = ode23('ex4eq2', 0, 0.0002, initial);
plot(t1, num1_i(:,1),t2, num2_i(:,1),'w',t3, num3_i(:,1)),'w',...
title('Inductor current for parallel RLC circuit with 3 resistor values'),...
xlabel('Time (sec)'),...
ylabel('I (amps)'),...
grid
```

FIGURE 3.14 Listing of the M-file for the parallel *RLC* circuit in Fig. 3.13.

requested to input values for the current source, the inductor, and the capacitor, as well as three resistor values. For each of the resistor values the function `ode23()` is called with the name of the M-file that contains the equations specifying the set of first-order differential equations. This file is in the `\matlab` subdirectory with the name `ex4eq2.m`, and is listed in Fig. 3.15. Note that because the M-file containing the set of equations uses data gathered from the M-file that generates the solution, you must declare this data "global."

```
% set of first-order differential equations for a parallel RLC circuit
function i_prime = ex4eq2(t, i)
i_prime(1) = i(2);
i_prime(2) = I_source/L_times_C - i(2)/R_times_C - i(1)/L_times_C;
```

FIGURE 3.15 Listing of the M-file that contains the set of first-order differential equations that describe the behavior of the circuit in Fig. 3.13.

You use the `plot()` function to display the three current traces for the three different resistor values, which represent the underdamped, overdamped, and critically damped responses. The input dialog for this M-file is shown in Fig. 3.16. The resulting plots are shown in Fig. 3.17. You can use this M-file to explore the effect of other resistor values on the inductor current in this circuit.

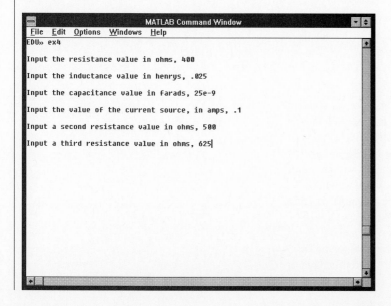

FIGURE 3.16 Input to the M-file listed in Fig. 3.15.

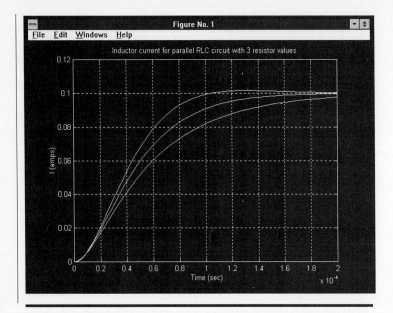

FIGURE 3.17 Plot of the inductor current for the three different resistor values in the circuit of Fig. 3.13.

EXAMPLE 3.5

Find the partial fraction expansion of the transfer function

$$H(s) = \frac{180(s + 30)}{s(s + 5)(s + 3)^2}.$$

Analytical techniques for finding partial fraction expansions are in Chapter 14 of the text.

SOLUTION

As you recall from the parent text, you can use partial fraction expansion of a transfer function (or any ratio of two polynomials in s) to find the time response of a circuit using the inverse Laplace transform. MATLAB makes it particularly easy to compute the partial fraction expansion of any ratio of two polynomials. The numerator and denominator polynomials are represented as vectors, where the components of the vectors are the coefficients of the polynomials, from the highest power to the constant term. Then you use the residue() function in MATLAB to generate the partial fraction expansion.

We begin by expanding the numerator and denominator of $H(s)$ into polynomial form, so that you can determine the coefficients of each power of s:

$$H(s) = \frac{180(s + 30)}{s(s + 5)(s + 3)^2} = \frac{180s + 5400}{s^4 + 11s^3 + 39s^2 + 45s + 0}.$$

The M-file that specifies these coefficients and uses the `residue()` function to compute the partial fraction expansion is found in the `\matlab` subdirectory with the filename `ex5.m`. It is also listed in Fig. 3.18.

```
num = [180 5400];
den = [1 11 39 45 0];
[r,p,k] = residue(num, den);
r
p
```

FIGURE 3.18 Listing of the M-file that computes the partial fraction expansion of the transfer function for Example 3.5.

The output from MATLAB is shown in Fig. 3.19 and requires a bit of interpretation. The `residue` function computes the coefficients of each term of the expansion as the vector r, and the roots for each term as the vector p. From Fig. 3.19, we see that the partial fraction expansion is

$$H(s) = \frac{-225}{s+5} + \frac{105}{s+3} - \frac{810}{(s+3)^2} + \frac{120}{s}.$$

The `residue()` function is an easy way to confirm your hand calculations of the partial fraction coefficients. Try using it on several other ratios of polynomials.

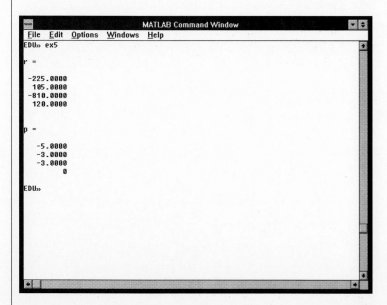

FIGURE 3.19 MATLAB computation of the partial fraction expansion of the transfer function in Example 3.5.

The transfer function for a second-order low-pass filter is given by

$$H(s) = \frac{K}{(s/\omega_c)^2 + 2\zeta s/\omega_c + 1},$$

where K is the dc gain, ω_c is the center frequency, and ζ is the damping ratio associated with the second-order characteristic. Construct the frequency response plots in Bode form for this transfer function with $K = 10$, $\omega_c = 1000$ rad/s, and three different damping ratios — 0.05, 0.3, and 0.7071. The text discusses techniques for constructing Bode plots by hand in Section 15.6.

S O L U T I O N

MATLAB provides a function that greatly simplifies the construction of Bode frequency response plots, thereby providing you with an easy method for visualizing the frequency response of a circuit whose transfer function you have computed. The MATLAB function `bode()` requires the transfer function to be specified as a numerator polynomial and a denominator polynomial, represented as vectors containing the coefficients of s. This method of representing a transfer function was illustrated in Example 3.5. The `bode()` function computes the magnitude and phase angle for a range of frequencies where there is variation in the frequency response. We can then use MATLAB to plot the results.

The M-file for this problem is in the subdirectory \matlab with the filename ex6.m. It is listed in Fig. 3.20. Note that the user inputs

```
zeta = input('Input the damping, ');
center = input('Input the center frequency, ');
gain = input('Input the dc gain, ');
K = gain;
a = 2*zeta/center;
b = 1/(center*center);
num = [K];
den = [b a 1];
[mag1, phase1, w1] = bode(num,den);
zeta = input('Input another damping, ');
a = 2*zeta/center;
b = 1/(center*center);
num = [K];
den = [b a 1];
```

(File continued on next page.)

```
[mag2, phase2, w2] = bode(num,den);
zeta = input('Input another damping, ');
a = 2*zeta/center;
b = 1/(center*center);
num = [K];
den = [b a 1];
[mag3, phase3, w3] = bode(num,den);
subplot(2,1,1), plot(log10(w1), 20*log10(mag1), 'w', log10(w2), 20*log10(mag2),
     'w', log10(w3),20*log10(mag3),'w')
xlabel('Log Frequency (rad/s)')
ylabel('Gain (dB)')
title('Bode magnitude plot of complex poles with 3 values of damping')
%pause
subplot(2,1,2), plot(log10(w1), phase1, 'w', log10(w2), phase2, 'w', log10(w3),
     phase3, 'w')
xlabel('Log Frequency (rad/s)')
ylabel('Phase (deg)')
title('Bode phase angle plot of complex poles with 3 values of damping')
```

FIGURE 3.20 Listing of the M-file for Example 3.6.

the values of the center frequency, damping ratio, and dc gain as prompted by MATLAB. The interaction with the user is shown in Fig. 3.21. The numerator and denominator polynomials are then constructed and used as input to the bode() function. This process is repeated for each of the three damping ratios. Magnitude in dB and phase angle in degrees are plotted on separate subplots versus the log of the frequency. The familiar result is shown in Fig. 3.22.

MATLAB provides a simple way to calculate and plot the frequency response plots of even complex circuits. Use this M-file as a template

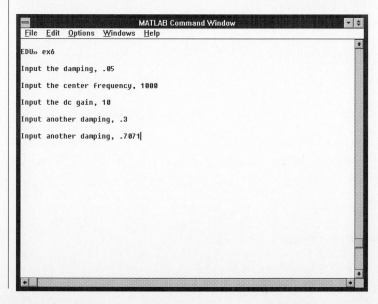

```
MATLAB Command Window
 File  Edit  Options  Windows  Help

EDU» ex6

Input the damping, .05

Input the center frequency, 1000

Input the dc gain, 10

Input another damping, .3

Input another damping, .7071
```

FIGURE 3.21 User input to the M-file shown in Fig. 3.20, specifying the parameters of a second-order low-pass filter.

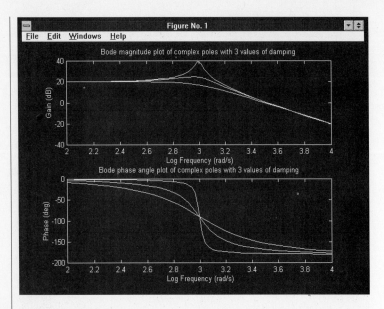

FIGURE 3.22 Bode plot for the second-order low-pass filter with varying damping ratio.

to develop the frequency response plot of some of the filters in Chapter 16 of the text. By varying the filter parameters, you can begin to understand the effect of each parameter on the frequency response characteristics of the circuit.

E X A M P L E   3.7

Use MATLAB to construct the sawtooth waveform in Fig. 3.23 with a Fourier series. Construct plots of the first harmonic, the sum of the first two harmonics, the sum of the first five harmonics, and the sum of the first ten harmonics. Fourier series expansions of periodic functions are discussed in Chapter 17 of the text.

S O L U T I O N

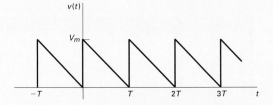

FIGURE 3.23 A sawtooth waveform for Example 3.7.

Example 17.1 in the parent text computes the Fourier series for one type of sawtooth waveform. Using the methods of Example 17.1, we can show that the Fourier series for the sawtooth in Fig. 3.23 is

$$v(t) = \frac{V_m}{2} + \frac{V_m}{\pi} \sin \omega_0 t + \frac{V_m}{2\pi} \sin 2\omega_0 t + \frac{V_m}{3\pi} \sin 3\omega_0 t + \cdots.$$

We can use the MATLAB $\texttt{sin()}$ function to construct the Fourier series with as many terms as we care to include.

The M-file that computes the Fourier series is in the subdirectory \matlab with the filename ex7.m. A listing of this M-file is seen in Fig. 3.24. From this M-file you can see we chose a fundamental

```
clc
%       The Fourier series expansion for a sawtooth-wave is made up of a sum
%       of harmonics.
%       The fundamental frequency
t = 0:.1:10;
y = 0.5 + sin(t)/pi;
subplot(2,2,1), plot(t,y,'w')
title('Sawtooth: first harmonic')
xlabel('t (sec)')
ylabel('V (volts)')
axis;
%       Add the next harmonic to the fundamental
y = 0.5 + sin(t)/pi + sin(2*t)/(2*pi);
subplot(2,2,2), plot(t,y,'w')
title('Sawtooth: first two harmonics')
xlabel('t (sec)')
ylabel('V (volts)')
%       The first five harmonics:
y1 = 0.5+sin(t)/pi+sin(2*t)/(2*pi)+sin(3*t)/(3*pi);
y = y1 + sin(4*t)/(4*pi)+sin(5*t)/(5*pi);
subplot(2,2,3), plot(t,y,'w')
title('Sawtooth: first five harmonics')
xlabel('t (sec)')
ylabel('V (volts)')
%       The first ten harmonics:
y1 = 0.5+sin(t)/pi+sin(2*t)/(2*pi)+sin(3*t)/(3*pi);
y2 = y1+sin(4*t)/(4*pi)+sin(5*t)/(5*pi)+sin(6*t)/(6*pi);
y3 = y2+sin(7*t)/(7*pi)+sin(8*t)/(8*pi)+sin(9*t)/(9*pi);
y = y3+sin(10*t)/(10*pi);
subplot(2,2,4), plot(t,y,'w')
title('Sawtooth: first ten harmonics')
xlabel('t (sec)')
ylabel('V (volts)')
pause
%       The fundamental to the 10th harmonic:
t = 0:.031:3.14;
y = zeros(10,max(size(t)));
x = zeros(size(t));
```

(File continued on next page.)

```
for k=1:1:10
x = x + sin(k*t)/(k*pi);
y(k,:) = .5 + x;
end
%       Plot this as a 3-d mesh surface
axis;
subplot(111), mesh(y)
title('3-d mesh surface of the transition to a sawtooth wave')
```

FIGURE 3.24 A listing of the M-file to generate a Fourier series for the sawtooth wave in Fig. 3.23.

frequency of 1 rad/s, which is equivalent to a period of about 6.3 seconds. We specify a time span of 0 to 10 seconds, so we can examine almost two periods of the waveform.

Now it is a simple matter to sum the terms in the Fourier series. We divide the screen into four subplots and use these four plotting "windows" to show a graph of the developing sawtooth waveform. The result is shown in Fig. 3.25. You might want to edit the M-file to look at different combinations of the harmonic components, or to see how many terms must be included before the sawtooth looks like the plot in Fig. 3.23. You can also use this M-file as a template to generate the Fourier series for other waveforms, such as the square-wave or half-wave rectified sinusoid.

Finally, to hint at the power MATLAB has to generate compelling visual images, we produce a three-dimensional mesh surface plot of the sum of the first ten terms in the Fourier series representation of the sawtooth wave. The result is shown in Fig. 3.26. Simply by changing the M-file, you can examine a similar three-dimensional plot for a greater or lesser number of terms in the Fourier series.

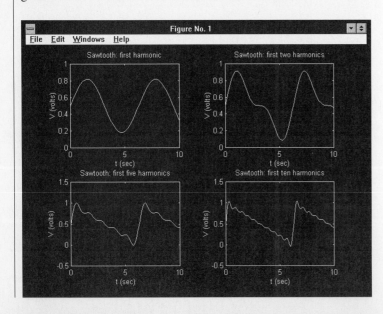

FIGURE 3.25 Building a sawtooth waveform from the Fourier series.

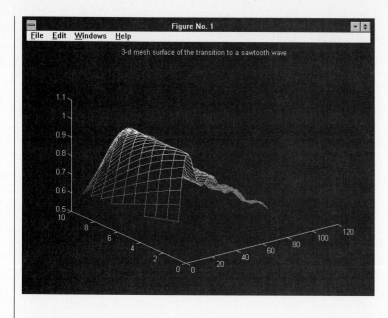

FIGURE 3.26 A three-dimensional plot of the first ten harmonic terms in the Fourier series for a sawtooth wave.

BIBLIOGRAPHY

D. M. Etter, *Engineering Problem Solving with MATLAB*, Prentice-Hall, Englewood Cliffs, NJ, 1993.

The MathWorks, Inc., *The Student Edition of MATLAB*, Prentice-Hall, Englewood Cliffs, NJ, 1992.

4 SPREADSHEETS

You might not have realized that spreadsheet programs can assist you in your study of linear circuits. After all, most of us think of spreadsheets as tools for balancing checkbooks, collecting and analyzing experimental data, or computing end-of-semester grades. Here we illustrate a few of the many uses for spreadsheet programs in the context of linear circuit analysis.

Spreadsheet programs excel at repetitive computations. That is, a spreadsheet is best at performing the same set of actions on different data values. How can you find such repetitive activities in circuit analysis? One example is a circuit whose topology stays the same, but whose component values vary. You can use the spreadsheet to analyze the topology for a range of component values, perhaps in search of a component value that maximizes or minimizes some parameter in the circuit. Another example is the time response of a circuit. Once you compute the time response in closed form, you can use the spreadsheet to compute the response for a range of time values. Since most spreadsheet programs provide a means for graphical output, you can use the spreadsheet to produce time-domain plots of circuit behavior.

In a similar manner, you can use spreadsheet programs to plot the frequency response of a circuit. The frequency response is simply a plot of the magnitude and phase angle of the circuit's output as the frequency is varied over some range of values. Repetitively calculating magnitude and phase angle for a large number of frequencies is an easy task for a spreadsheet program.

Therefore, any circuit analysis that involves repetitive computation for a range of circuit parameter values is a candidate for assistance using a spreadsheet. Clearly, spreadsheet programs were not designed with the primary goal of supporting circuit analysis. Therefore, in what follows, we will employ only some of the many capabilities of a typical spreadsheet program. We assume that you have consulted a reference to learn the fundamentals about your spreadsheet program. The examples below assume you know the basics of how to use a spreadsheet, and are intended to help you set up a circuit analysis problem in a form suitable for analysis by a spreadsheet.

In order to use a spreadsheet program to help in circuit analysis, you must usually perform at least some of the preliminary assessment of the circuit. This is because the spreadsheet works with numbers and formulas, not with circuit schematics. But often, the spreadsheet can help you compute intermediate results in the same order in which you would compute them if working the problem by hand. These intermediate results can help you to learn more about the performance of the circuit as a whole, and you may use them to confirm or correct the solutions you generated previously by hand. The trick in using a spreadsheet program to assist in circuit analysis is to let the spreadsheet calculate while you perform the analysis. We will illustrate this idea in the several examples below.

We use Quattro® Pro in the examples that follow, and employ only a small subset of the many useful tools in this powerful spreadsheet package.[1] We restrict our attention to fairly simple functional forms, and make use of the graphing features to construct a visual interface to the data. We do not use any of the high-level analysis tools provided in QuattroPro, even though some of them might be very useful in circuit analysis, because these tools might not be available for your spreadsheet. You can adapt quite easily most of the techniques we use in the examples for any of the other common spreadsheet programs.

---

[1] The examples in this chapter were generated using QuattroPro for Windows, Version 5.0.

EXAMPLE 4.1

Find the equivalent resistance R_{ab} for the circuit in Fig. 4.1, using several different values for R. This circuit has the same topology as Problem 3.47 in the parent text.

SOLUTION

We begin by filling the first column with candidate values for R. For this example, we use $9, 1, 18, 50$, and $9000\ \Omega$. Feel free to choose these or any other values of interest. Next, we could employ the many circuit simplification techniques to arrive at the formula that calculates R_{ab} from any value of R. But instead, we let the spreadsheet do the intermediate calculations as we simplify the circuit step by step.

Looking at the circuit in Fig. 4.1, we see that we can begin by simplifying the topmost Δ, recognizing that its resistance is equivalent to $2R \| R$. In the spreadsheet, if the column that contains the data is named A, the data in the second column will be the result of this first simplification by specifying the function $2A * A/(2A + A)$. This first step is shown in Fig. 4.2. Note that we have labeled the columns in two ways — with a name, which in Column B is "Step 1," and below that with the formula used to calculate the data in this column. This will allow you to follow the computations as we construct them. You can find this example in the file named \qpro\ex1.wb1.

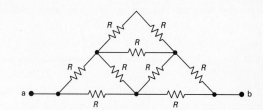

FIGURE 4.1 The circuit for Example 4.1.

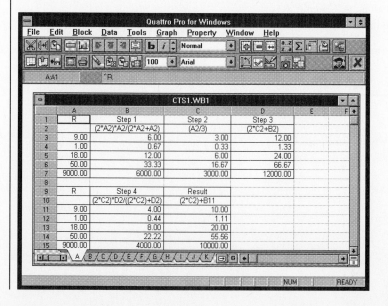

FIGURE 4.2 The spreadsheet computations that simplify the circuit in Fig. 4.1 to a single resistance.

The next step in the simplification works with the two sets of Δ-connected resistors on the left and right sides of the circuit in Fig. 4.1. Transforming the Δ connection to the equivalent Y connection will yield the circuit shown in Fig. 4.3. Since the resistors in the Δ all have the value R, the resistors in the equivalent Y will all have the value $R/3$. As you can see from the spreadsheet in Fig. 4.2, the next column, C, is used to compute the values of the Y-connected resistors.

The next step in simplification reduces the upper three series-connected resistors in Fig. 4.3 to a single equivalent resistor by adding the resistances. Recall that Column B contains the value of the middle resistor (the parallel combination of $2R$ and R), while Column C contains the value of the other two resistors ($R/3$). You can compute the equivalent resistor value using the formula $2*C+B$, as shown in Column D in Fig. 4.2.

The next step in simplification is to combine the equivalent resistor from the last step in parallel with the middle two resistors in the lower branch of Fig. 4.3. The equivalent resistor is in Column D, while the value of each of the two middle resistors is in Column C. Thus, in parallel, the formula is $2C\|D = 2C*D/(2C+D)$. Although you would probably use Column E for this formula, we have used the lower rows of Column A, so that all the spreadsheet data would fit on a single screen image, as you can see in Fig. 4.2.

The final step is to combine the equivalent resistor from the previous step in series with the two remaining resistors in the lower branch of the circuit. This step is achieved in the column labeled "Result." As you can see, if all of the resistors have the value 9 Ω, the equivalent resistance of this complex circuit is 10 Ω. Feel free to experiment with other resistor values, either by replacing values in Column A with ones of interest to you, or by adding more values.

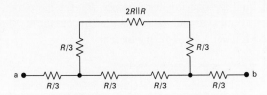

FIGURE 4.3 An intermediate step in reducing the circuit in Fig. 4.1 to an equivalent resistance.

EXAMPLE 4.2

For the circuit shown in Fig. 4.4, vary the load resistor and plot the power absorbed by the load as a function of load resistance. This is Problem 4.73 from the parent text.

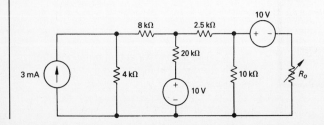

FIGURE 4.4 The circuit for Example 4.2.

SOLUTION

In order to find the power absorbed by the load resistor, it is easiest to determine the current through the load resistor, i. Then the power absorbed is given by $P = Ri^2$. Chapter 4 of the parent text provides the background needed to write and solve the three mesh equations that describe this circuit. The result of mesh analysis gives

$$i = \frac{-4.375}{5000 + R_o}.$$

Circuit analysis has yielded the formulas we need to have the spreadsheet compute power as a function of varying resistance. You can find the spreadsheet in the file \qpro\ex2.wb1. The screen image of the spreadsheet is shown in Fig. 4.5. The first column contains 10 load resistor values, from 1 kΩ to 10 kΩ. The second column computes the current through the load resistor, using the formula $-4.375/(5000 + A * 1000)$. The third column calculates the power absorbed by the load resistor, using the formula $A * B * B$. The final column expresses the power in microwatts by multiplying the previous column by 10^6.

While you might be able to learn something about the relationship between load resistance and power absorbed by staring at the numbers in the final column, it is much easier to view this relationship in the form of a plot. We need only plot the numbers in the last column versus the numbers in the first column. The result is shown in Fig. 4.6. We have added a title and axis labels to the plot to make it more meaningful. From this plot it is easy to see that the power is a maximum

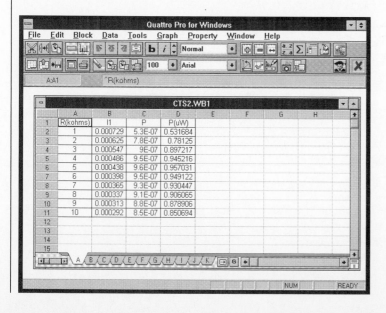

FIGURE 4.5 The spreadsheet analysis of load power as a function of load resistance for the circuit in Fig. 4.4.

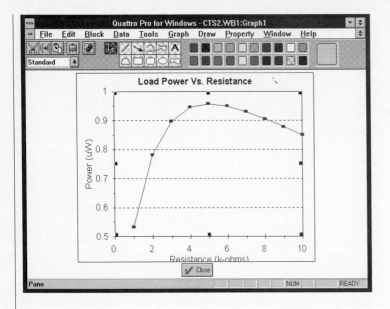

FIGURE 4.6 A plot of load power versus load resistance for the circuit in Fig. 4.4.

between a load resistance of 4 kΩ and 6 kΩ. You might want to choose a new set of load resistor values within this range, recalculate the power absorbed, and plot this new range of load resistance to make a better estimate of the load resistance that results in maximum power transferred.

EXAMPLE 4.3

The circuit shown in Fig. 4.7 is operating at a frequency of 10 rad/s. Assume $0 \leq \sigma \leq 10$. Plot the real and imaginary part of Z_{ab}. Is there a value of σ for which Z_{ab} is purely resistive? This example is adapted from Problem 9.59 in the text.

SOLUTION

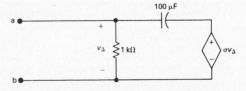

FIGURE 4.7 The circuit for Example 4.3.

Techniques presented in Chapter 9 of the parent text enable us to determine that the Thévenin equivalent impedance from the a,b terminals is given by

$$Z_{TH} = Z_{ab} = \frac{1000}{1 + j(1 - \sigma)} = \frac{1000}{1 + (1 - \sigma)^2} - j\frac{1000(1 - \sigma)}{1 + (1 - \sigma)^2}.$$

It is easy to use a spreadsheet to calculate the real and imaginary parts of this impedance for the range of σ values. You can find the spreadsheet in the file \qpro\ex3.wb1. Part of the spreadsheet is shown in Fig. 4.8.

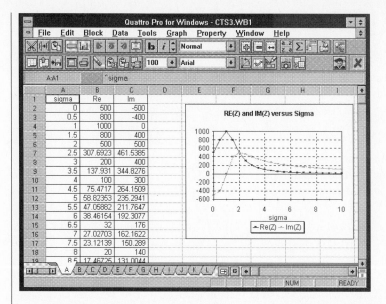

FIGURE 4.8 Spreadsheet computation of the real and imaginary parts of the impedance in the circuit in Fig 4.7 as σ is varied.

As before, looking at the formula for the real and imaginary part of the impedance, or even looking at the calculated values as σ is varied, is of limited use. But when we generate a plot of the real and imaginary parts of the impedance versus σ, shown in Fig. 4.8, the relationship between σ and Z_{ab} becomes much clearer. From the plot of the imaginary part of the impedance, we see that initially, the imaginary part is negative, so the equivalent impedance is capacitive. As σ increases, the imaginary part of the impedance decreases (and the real part of the impedance increases). At $\sigma = 1$, the imaginary part of the impedance is zero, the equivalent impedance is purely resistive, and the real part of the impedance is a maximum. As σ continues to increase, the imaginary part of the impedance increases, peaks for $\sigma = 2$, and then decreases, but is always positive. Thus, for $\sigma < 2$ the equivalent impedance is capacitive, for $\sigma = 0$ the equivalent impedance is purely resistive, and for $\sigma > 2$ the equivalent impedance is inductive.

You might want to explore the effect on equivalent impedance of a different range of σ values.

EXAMPLE 4.4

Make an amplitude plot for the transfer function I_o/I_g in the circuit shown in Fig. 4.9. This circuit is adapted from Problem 15.32 in the text.

SOLUTION

We can compute the transfer function by transforming the circuit components to their s-domain values, using current division, and simplifying. The result is

$$H(s) = \frac{s(s + 5000)}{s^2 + 5000s + 4 \times 10^8} .$$

To calculate the magnitude of $H(s)$ as a function of frequency, we make the substitution $s = j\omega$, find the magnitude of the numerator and denominator, and take their ratio. The magnitude of the numerator is

$$\sqrt{\omega^4 + (5000\omega)^2}$$

while the magnitude of the denominator is

$$\sqrt{(4 \times 10^8 - \omega^2)^2 + (5000\omega)^2}.$$

We use the spreadsheet to specify a range of frequencies, from 1000 rad/s to 100,000 rad/sec, and to compute the magnitude of the numerator and denominator at each of these frequencies, as shown in Fig. 4.10. You can find the spreadsheet in the file \qpro\ex4.wb1. Note that only a small subset of the range of frequencies and computed values fits on the screen shown in Fig. 4.10.

Next, we take the ratio of the numerator and denominator magnitudes to calculate $|H(j\omega)|$, as shown in the column so labeled in

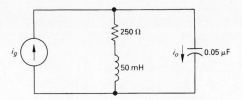

FIGURE 4.9 The circuit for Example 4.4.

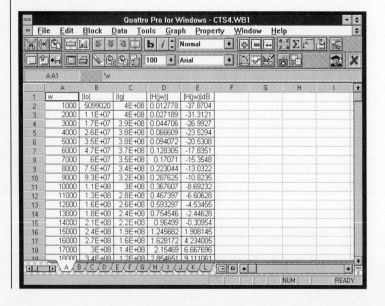

FIGURE 4.10 Spreadsheet computations for the magnitude of the transfer function of the circuit in Fig. 4.9 versus frequency.

Fig. 4.10. Finally, in order to generate a plot in Bode form, we calculate the magnitude in dB, which appears in the last column in the spreadsheet. The plot is shown in Fig. 4.11. Note that we used a logarithmic horizontal scale and a linear vertical scale, and that we titled the plot and labeled the axes for added clarity. From this plot we estimate that the center frequency is 20,000 rad/s. You might want to calculate the magnitude values for a small range of values about this estimated center frequency to make a more accurate assessment of the actual center frequency, as well as to enable a calculation of the bandwidth.

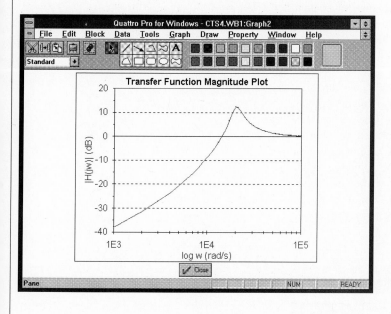

FIGURE 4.11 Bode magnitude plot for the circuit in Fig. 4.9.

EXAMPLE 4.5

Use the spreadsheet to calculate the first 10 harmonic terms in the Fourier series representation of the sawtooth wave in Fig. 4.12. Plot the Fourier series representation for the sum of one, four, and eight terms.

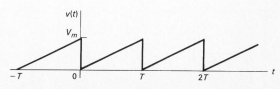

FIGURE 4.12 The sawtooth waveform for Example 4.5.

SOLUTION

The Fourier series representation of this sawtooth wave is determined in Example 17.1 of the parent text. The result is given as

$$v(t) = \frac{V_m}{2} - \frac{V_m}{\pi} \sum_{n=1}^{\infty} \frac{1}{n} \sin n\omega_0 t.$$

Let's choose $V_m = 1\text{V}$ and $\omega_0 = 1$ rad/sec. This will give us a period of about 6.3 seconds. The first column will contain the values of t, the second column will contain the dc term, the third column will contain the sum of the dc and the first harmonic, the fourth column will contain the sum of the dc, first, and second harmonics, and so forth. We use two functions, @sin and @pi, in constructing the Fourier terms. You can find the spreadsheet in the file \qpro\ex5.wb1. Part of it is shown in Fig. 4.13.

This computed data is rather overwhelming, and fairly meaningless, in spreadsheet form. But when we plot a few of the columns versus time, we begin to see the sawtooth wave emerge. One such plot is shown in Fig. 4.14. As you can see, we have plotted the sum of the dc and first harmonic terms, the sum of the dc and first through fourth harmonic terms, and the sum of the dc and first through eighth harmonic terms. You should try other plots of the data to get a better feel for how the sawtooth wave emerges from the sum of sinusoids. You can also use this spreadsheet as a template to generate the Fourier series for other periodic waveforms.

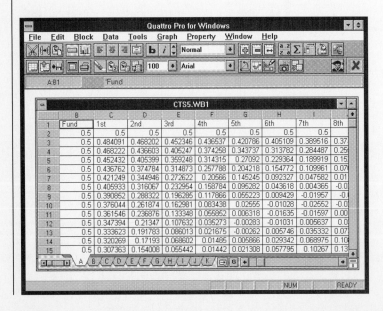

FIGURE 4.13 Spreadsheet computation of the terms in the Fourier series for the sawtooth wave in Fig. 4.12.

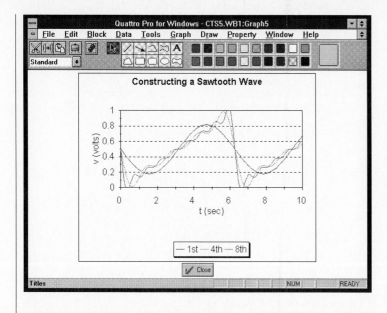

FIGURE 4.14 Plot of the Fourier series representation of a sawtooth wave, using different numbers of terms in the series.

BIBLIOGRAPHY

Borland, *QuattroPro for Windows, Version 5.0, User's Guide,* Borland International, 1993.

5 SYMBOLIC EQUATION SOLVERS

A symbolic equation solver is a computer tool that can help you deal with complicated mathematical problems. The idea of a computer program that can solve college-level calculus problems was first proposed by the artificial intelligence community as a suitable application to demonstrate the power of artificial intelligence principles. The key notion in these software tools is that they deal with symbols, as well as with numbers. There is often a need to solve a differential equation or an integral equation that has general, symbolic coefficients rather than numeric coefficients, and to construct a solution that is a function of these symbolic coefficients.

For example, consider the equation for the voltage across an inductor:

$$v(t) = L\frac{di}{dt}.$$

Suppose we want to determine the equation for the current through an inductor, for any value of inductance. Using elementary calculus, we can integrate both sides of the equation and arrive at the result

$$i(t) = \frac{1}{L} \int_{t_0}^{t} v(\tau)\, d\tau + i(t_0).$$

This is what a symbolic equation solver tool does, among other things. Whereas most software packages that perform integration must use numerical methods, and therefore must restrict the integration to equations that are free of symbols, the symbolic equation solver can integrate without using numerical methods, so is free of the requirement that the equations have no symbols.

How can a symbolic equation solver help you in linear circuit analysis? There are many possible applications. Since linear circuits are modeled mathematically by sets of algebraic equations (in the case of purely resistive circuits) or sets of differential equations, a symbolic equation solver can produce general solutions to these sets of equations, or specific solutions if numerical values are provided. You can use these solutions to confirm the solutions you generate by hand, or solve higher-order circuits which are too cumbersome to generate by hand. In addition, most symbolic equation solvers have a graphical interface, which enables you to visualize solutions and gain a deeper understanding of the effect of parameter variation on circuit behavior.

The particular symbolic equation solver used for the examples is Maple V (Waterloo).[1] It is assumed that you have learned the Maple basics in another course, or have used a reference text or the extensive help screens and on-line tutorial provided in Maple to learn it on your own. The examples make use of only a small portion of the total capabilities of Maple, as we apply this powerful software tool to the study of linear circuits. Each of the examples exists as a worksheet on the disk, so you can follow along with the text.

As you will see, you will need to express your circuit analysis problem in the form of an equation or a set of equations in order to use Maple to assist you in solving the circuit. Maple has a particular syntax used to describe the equations in a form that it understands, but that is a bit hard to read. But Maple will echo back the equations you specify in a style called "typographic," which is much easier to read and understand.

---

[1] The examples in this chapter were generated using Maple V for Windows, Version 3.0.

EXAMPLE 5.1

Find the current i_1 in the circuit shown in Fig. 5.1. Assume that $V_1 = 240$ V, $R_1 = 1\ \Omega$, $R_2 = 5\ \Omega$, and $R_3 = 2\ \Omega$. This circuit is taken from Problem 4.98 in the text.

SOLUTION

You can find the worksheet for this example in the file `\maple` `\ex1.ms`. To begin, we must construct the set of five mesh equations that describe the behavior of this circuit.

They are

$$240 = 4i_a - i_b - i_c - i_d - i_e;$$
$$0 = -i_a + 8i_b - i_c;$$
$$0 = -i_a - i_b + 5i_c - i_d;$$
$$0 = -i_a - i_c + 5i_d - i_e;$$
$$0 = -i_a - i_d + 8i_e.$$

We can use the `solve()` function to describe these equations to Maple and have it generate a solution. This portion of the worksheet is shown in Fig. 5.2. Note that the description of the set of equations

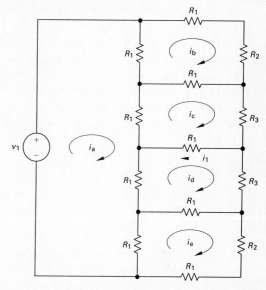

FIGURE 5.1 The circuit for Example 5.1.

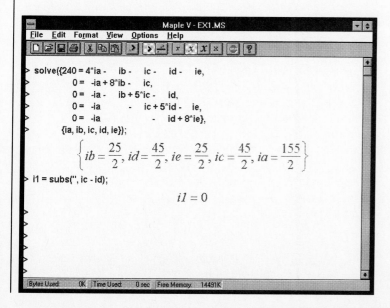

FIGURE 5.2 The Maple solution of the five describing equations for the circuit in Fig. 5.1.

need not fit on one line, and equations are separated by commas. Maple waits to analyze your input until it sees a semicolon.

Maple solves for all five of the mesh currents, as shown in Fig. 5.2. Since $i_1 = i_c - i_d$, we find that the value of i_1 is 0 amps.

Now, let's ask Maple to solve these equations in general. That is, instead of using specific values for the resistors, let's use symbols such as R_1, R_2, and R_3, as shown in Fig. 5.1. This time, we have to develop the set of equations employing the names of the resistors, rather than their values. The describing set of equations is

$$V_1 = 4R_1i_a - R_1i_b - R_1i_c - R_1i_d - R_1i_e;$$
$$0 = -R_1i_a + (3R_1 + R_2)i_b - R_1i_c;$$
$$0 = -R_1i_a - R_1i_b + (3R_1 + R_3)i_c - R_1i_d;$$
$$0 = -R_1i_a - R_1i_c + (3R_1 + R_3)i_d - R_1i_e;$$
$$0 = -R_1i_a - R_1i_d + (3R_1 + R_2)i_e.$$

We describe this set of equations to Maple and ask for their solution using `solve()`, just as before. This time, Maple responds with symbolic solutions for the five mesh currents. This part of the worksheet is shown in Fig. 5.3. You can use these symbolic solutions to calculate the mesh currents for any combination of resistor values. Note from these symbolic solutions that the choice of resistor values does not affect the value of the current i_1. Since $i_1 = i_c - i_d$, and we see from Maple's solution that $i_c = i_d$, the value of i_1 is always 0 amps, no matter what the value of the resistors.

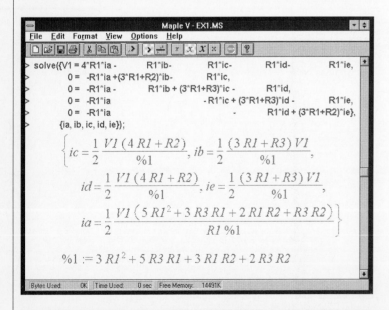

FIGURE 5.3 Maple solution of the five describing equation for the circuit in Fig. 5.1 for any values of resistance.

EXAMPLE 5.2

For the circuit in Fig. 5.4, plot the current through the inductor, compute and plot the voltage across the inductor, compute and plot the power in the inductor, and compute and plot the energy stored by the inductor. This analysis is performed in Example 6.3 in the parent text.

SOLUTION

You can find the worksheet for this example in the file \maple \ex2.ms. We begin by specifying and plotting the current supplied by the current source. We need the exp() function to specify the current, and the plot() function to plot it. The relevant portion of the Maple worksheet is shown in Fig. 5.5, and the plot of the current versus time is shown in Fig. 5.6.

Next, we describe the differential equation that relates voltage and current in an inductor, using the diff() function. Maple then computes the voltage. Note that we have used a symbol to represent the inductor value, so that the voltage computed by Maple is correct for any inductor. If we specify an inductor value, such as 10 H, we can then generate a plot of voltage versus time for the inductor. This section of the Maple worksheet is shown in Fig. 5.7, and the resulting plot of voltage across the inductor is shown in Fig. 5.8.

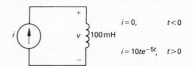

$$i = 0, \qquad t < 0$$
$$i = 10te^{-5t}, \quad t > 0$$

FIGURE 5.4 The circuit for Example 5.2.

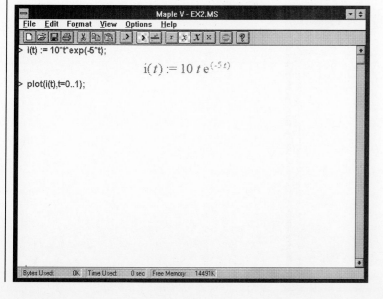

FIGURE 5.5 The Maple representation of the current source in Fig. 5.4.

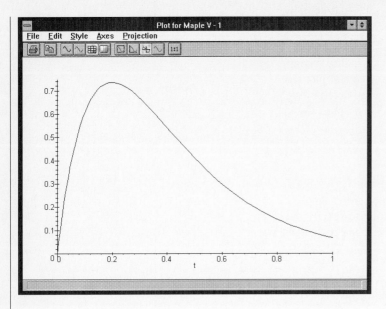

FIGURE 5.6 A plot of the current through the inductor in Fig. 5.4.

Next, we describe the equation that calculates the power in the inductor, and Maple solves that equation using all that we have described previously. This portion of the Maple worksheet is shown in Fig. 5.9. We can plot power versus time, as shown in Fig. 5.10. From the plot we see that for the first 0.2 seconds, energy is being stored in the inductor, while for the remaining time, energy is being extracted from the inductor.

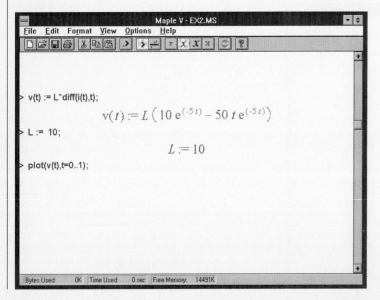

FIGURE 5.7 The Maple solution for the voltage across the inductor in Fig. 5.4.

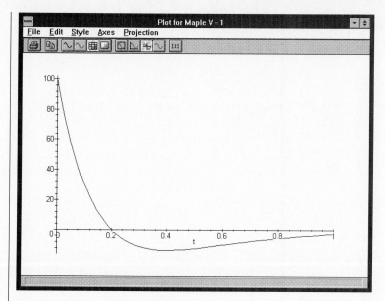

FIGURE 5.8 A plot of the voltage across the inductor in Fig. 5.4.

The total area under the power curve gives us the total energy stored in or extracted from the inductor as $t \to \infty$. Maple can compute that area by using the int () function to integrate the equation for power from $t = 0$ to ∞. We see in Fig. 5.9 that the total energy is zero, as it should be for this ideal inductor.

Finally, we can calculate the expression for the energy in the inductor by describing this equation to Maple. Maple expresses the energy equation using all of the information previously provided. The result

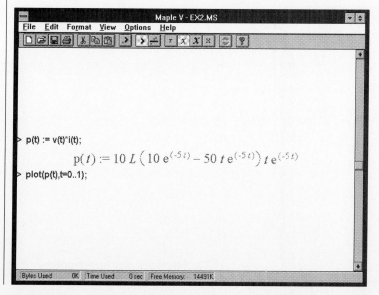

FIGURE 5.9 The Maple solution for the power in the inductor in Fig. 5.4.

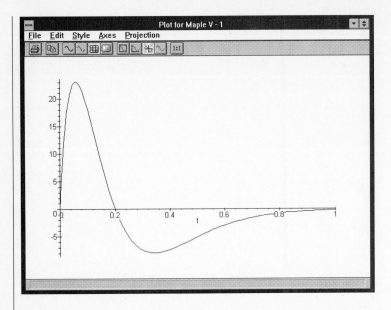

FIGURE 5.10 A plot of the power in the inductor in Fig. 5.4.

is shown in Fig. 5.11. We can plot this energy equation as a function of time, as shown in Fig. 5.12. Note that as $t \to \infty$, the energy goes to 0, as Maple calculated previously.

You can use this worksheet as a template to explore other forms of source current and their effect on the voltage, power, and energy for an inductor. You might also replace the inductor and current source with a capacitor and voltage source, to compare the behavior of capacitors with the behavior of inductors.

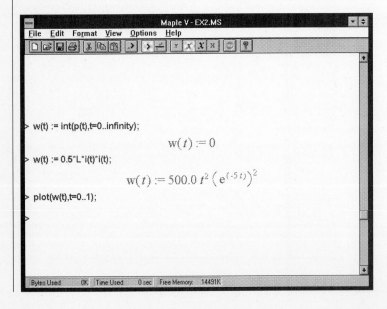

FIGURE 5.11 The Maple solution for the energy stored in the inductor in Fig. 5.4.

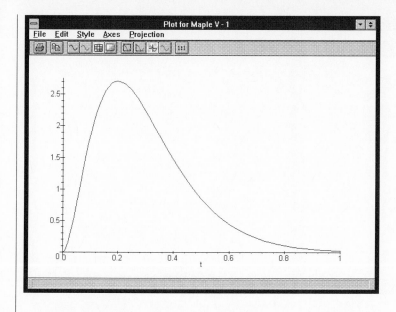

FIGURE 5.12 A plot of the energy stored in the inductor in Fig. 5.4.

EXAMPLE 5.3

Find the partial fraction expansion of the transfer function

$$H(s) = \frac{100(s + 3)}{s^3 + 12s^2 + 61s + 150}.$$

Partial fraction expansion is described in Chapter 14 of the text.

SOLUTION

You can find the worksheet for this example in the file `\maple\ex3.ms`. We begin by asking Maple to factor the denominator polynomial, so that we know the roots, or poles, of the transfer function. To do this, describe the polynomial to Maple and use the `factor()` function, as shown in the worksheet in Fig. 5.13.

Note from Fig. 5.13 that our initial attempt to factor the denominator polynomial did not factor the terms where the poles are complex. To extract the complex poles, we need to specify the term that we want to factor over the complex numbers. You can do this by including `I` as a second argument to `factor()`. Note that in Maple, `I` is a global symbol whose value is $\sqrt{-1}$, more familiar as the symbol j in electrical engineering. If you prefer to use the symbol j, you can do so by typing the Maple command `alias(I=I, j=sqrt(-1))`.

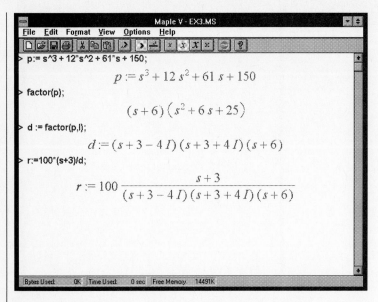

FIGURE 5.13 Maple factors the denominator of the transfer function.

As you can see from Fig. 5.13, when Maple factors over the complex numbers it finds the two complex roots, along with the single real root it found before. Using this factorization, we define the transfer function for Maple, which has the symbolic name r.

Now we can use the residue() function to compute the coefficients of the partial fractions. Maple must fetch this function from its library, which we specify using the readlib() function. The result of Maple's computation is shown in Fig. 5.14.

You can use this worksheet as a sample from which to generate the partial fraction expansion of any transfer function. Try it!

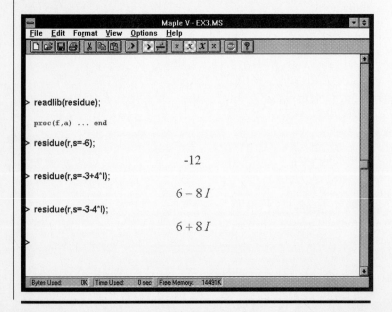

FIGURE 5.14 Maple calculates the residue for each of the poles in the transfer function.

EXAMPLE 5.4

Plot the magnitude of the transfer function $H(s) = I_o/I_g$ versus frequency for the circuit in Fig. 5.15. Frequency response plots are discussed in Chapter 14 of the text.

SOLUTION

We begin by deriving the transfer function for this circuit, following the methods described in Chapter 14 of the parent text. The result is

$$H(s) = \frac{s(s + 5000)}{s^2 + 5000s + 4 \times 10^8}.$$

To find the magnitude as a function of frequency, make the substitution $s = j\omega$, and calculate the form for the magnitude of the numerator and the denominator. We see that

$$|H(j\omega)| = \frac{\sqrt{\omega^4 + (5000\omega)^2}}{\sqrt{(4 \times 10^8 - \omega^2)^2 + (5000\omega)^2}}.$$

Now we can describe this equation for the transfer function magnitude to Maple. You can find the worksheet for this example in the file \maple\ex4.ms. The worksheet is shown in Fig. 5.16. The plot of the magnitude versus frequency is shown in Fig. 5.17. As you can see, this circuit is a bandpass filter with a center frequency of about 20,000 rad/s and a fairly narrow bandwidth.

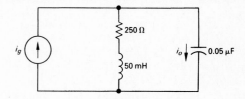

FIGURE 5.15 The circuit for Example 5.4.

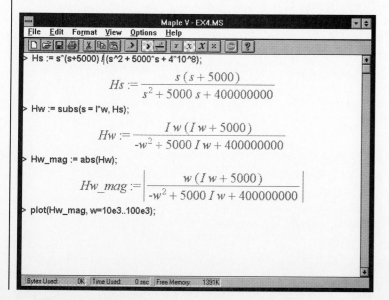

FIGURE 5.16 Describing the magnitude of the transfer function for the circuit in Fig. 5.15.

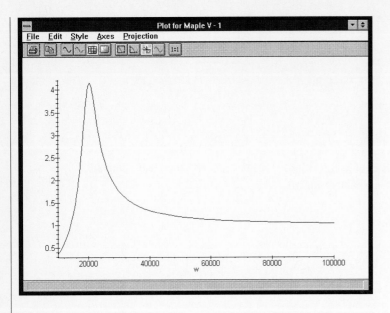

FIGURE 5.17 Plot of the transfer function magnitude versus frequency for the circuit in Fig. 5.15.

You can now use Maple to plot the phase angle of this transfer function versus frequency, or to explore the frequency response of any other circuit of interest.

EXAMPLE 5.5

Generate the sawtooth wave shown in Fig. 5.18 by using the first few terms of its Fourier series representation.

SOLUTION

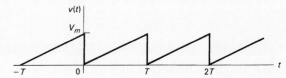

FIGURE 5.18 The sawtooth wave for Example 5.5.

The Fourier series for the sawtooth wave in Fig. 5.18 is derived in Example 17.1 in the parent text, and is given by

$$v(t) = \frac{V_m}{2} - \frac{V_m}{\pi} \sum_{n=1}^{\infty} \frac{1}{n} \sin n\omega_0 t.$$

Let's choose $V_m = 1$ V and $\omega_0 = 1$ rad/s (so $T = 2\pi$ seconds).

You can find the worksheet for this example in the file \maple \ex5.ms. The worksheet is shown in Fig. 5.19. Note that we use the sum() function to specify as many terms in the summation as we want. In this case, we specify 10 terms, using the second argument to the sum() function. Maple prints the summation in long form, as you can see in Fig. 5.19.

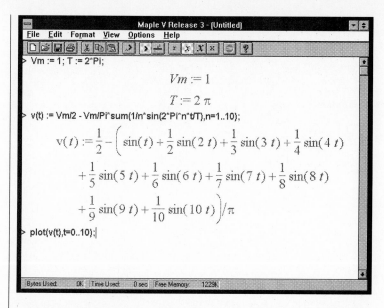

FIGURE 5.19 The Fourier series representation of the sawtooth wave in Fig. 5.18.

We can plot this approximation to the sawtooth wave, which contains 10 harmonic terms in the Fourier series. Since the period is about 6.3 seconds, we plot for 10 seconds. The plot is shown in Fig. 5.20.

You can look at the effect of adding more terms to the summation on the quality of the sawtooth wave by modifying the second argument to the sum() function. You can also use this worksheet as a template to explore the Fourier series representations of other periodic waveforms.

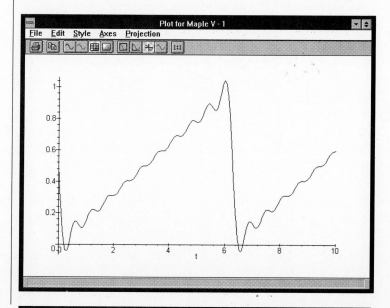

FIGURE 5.20 Plot of the Fourier series approximation of the sawtooth wave in Fig. 5.18.

BIBLIOGRAPHY

Bruce W. Char et al., *First Leaves: A Tutorial Introduction to Maple V*, Springer-Verlag, NY, 1992.

Bruce W. Char et al., *Maple V Language Reference Manual*, Springer-Verlag, NY, 1991.

Bruce W. Char et al., *Maple V Library Reference Manual*, Springer-Verlag, NY, 1991.

PROBLEMS

Here are nearly 150 problems modified from the parent text for you to solve using computer tools. You can find solutions to about 40 of the problems on the disk, organized by problem number. You will need to decide which of the computer tools available to you is the most appropriate for solving each problem. The number in brackets next to the problem number refers to the problem number in *Electric Circuits, 5/e*.

1. [1.3] The following data relates to generating motion picture film using computer graphics:

   * High-resolution film recorders have a resolution of 1200 × 1600 picture elements (pixels) per frame;

   * Each pixel requires 10 bits of data for each of the three primary colors—red, green, and blue;

   * It takes 10 floating-point calculations to determine the value of each color per pixel;

   * A motion picture runs 24 frames per second;

   * A supercomputer can perform 225 million floating-point calculations per second.

   Using this data, calculate the following:

   a) The time it takes to generate 1 s of film;

   b) The number of bytes of information that must be generated for 1 s of film (a byte is eight bits);

   c) The number of floppy disks needed to store the data for 1 s of film (a floppy disk can hold 1.2×2^{20} bytes of data).

2. [1.12] A 12 V battery supplies 100 mA to a radio. Plot the energy supplied by the battery over 4 hours.

3. [1.15] The voltage and current at the terminals of the circuit element in Fig. P3 are zero for $t < 0$. For $t \geq 0$ they are

$$v = 80{,}000te^{-500t} \text{ V}, \quad t \geq 0;$$

$$i = 15te^{-500t} \text{ A}, \quad t \geq 0.$$

a) Plot the power delivered to the circuit element for the first 10 ms.

b) From your plot, find the time (in milliseconds) when the power delivered to the circuit element is maximum, and the maximum value of p in milliwatts.

c) Plot the energy delivered to the circuit element for the first 10 ms.

d) From your plot, find the energy delivered to the circuit element for the first 10 ms, in microjoules.

e) Calculate the total energy delivered to the circuit element in microjoules.

FIGURE P3

4. [1.17] The voltage and current at the terminals of the circuit element in Fig. P4 are zero for $t < 0$. For $t \geq 0$ they are

$$v = (16{,}000t + 20)e^{-800t} \text{ V}$$

and

$$i = (128t + 0.16)e^{-800t} \text{ mA}.$$

a) Plot the voltage, current, and power delivered to the circuit for the first 10 ms.

b) From your plot, determine the instant in time at which the power is maximum, and the maximum power at that instant in watts.

c) Calculate the total energy delivered to the element, in millijoules.

FIGURE P4

5. [1.21] The voltage and current at the terminals of the element in Fig. P5 are

$$v = 125 \cos 400 \pi t \text{ V};$$

$$i = 32 \sin 400 \pi t \text{ A}.$$

a) Plot the voltage, current, and power for the element for the first 10 ms.

b) From your plot, find the maximum value of the power being extracted from the element and the maximum value of the power delivered to the element.

c) Plot the average value of p over the first 10 ms.

d) Find the average value of p in the interval $0 \leq t \leq 5$ ms.

e) Find the average value of p in the interval $0 \leq t \leq 6.25$ ms.

FIGURE P5

6. [1.23] The manufacturer of a 1.5 V D-cell flashlight battery says that the battery will deliver 9 mA for 40 continuous hours. During that time the voltage will drop from 1.5 V to 1.0 V. Assume that the drop in voltage is linear with time. Plot the energy delivered by the battery over the 40 h time interval. How much energy does the battery deliver in this 40 h interval?

7. [1.25] One method of checking calculations involving interconnected circuit elements is to see that the total power delivered equals the total power absorbed (conservation-of-energy principle). With this thought in mind, check the interconnection in Fig. P7 and state whether it satisfies this power check. The following table contains the current and voltage values for each element.

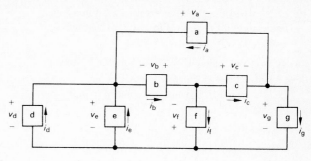

FIGURE P7

TABLE FOR 7

ELEMENT	VOLTAGE (V)	CURRENT (A)
a	160	−10
b	−100	20
c	60	6
d	80	50
e	800	−20
f	−700	14
g	640	16

8. [2.3] The current i_o in the circuit in Fig. P8 is 14 A. Find (a) i_a, (b) i_g, and (c) the power delivered by the independent current source.

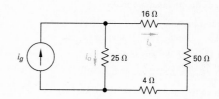

FIGURE P8

9. [2.12] The current i_o in the circuit in Fig. P9 is 4 A.

a) Find i_1.

b) Find the power dissipated in each resistor.

c) Verify that the total power dissipated in the circuit equals the power developed by the 180 V source.

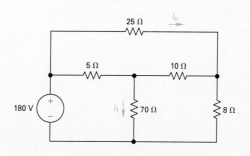

FIGURE P9

10. [2.16] The voltage and current were measured at the terminals of the device in Fig. P10(a). The results are tabulated in Fig. P10(b).

a) Construct a circuit model for this device, using an ideal voltage source and a resistor.

b) Use the model to predict the value of i_t when v_t is zero.

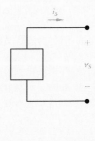

v_t (V)	i_t (A)
50	0
58	2
66	4
74	6
82	8
90	10

(a)   (b)

FIGURE P10

11. [2.18] The table in Fig. P11(a) shows the relationship between the terminal voltage and current of the practical constant-voltage source shown in Fig. P11(b).

a) Plot v_s versus i_s.

b) Construct a circuit model of the practical source that is valid for $0 \le i_s \le 24$ A, based on the equation of the line plotted in (a). (Use an ideal voltage source in series with an ideal resistor.)

c) Use your circuit model to predict the current delivered to a 1 Ω resistor connected to the terminals of the practical source.

d) Use your circuit model to predict the current delivered to a short-circuit connected to the terminals of the practical source.

e) What is the actual short-circuit current?

f) Explain why the answers to parts (d) and (e) are not the same.

v_s (V)	i_s (A)
24	0
22	8
20	16
18	24
15	32
10	40
0	48

(a)   (b)

FIGURE P11

12. [2.20] The variable resistor R in the circuit in Fig. P12 is adjusted until v_a equals 60 V. Find the value of R.

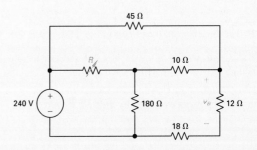

FIGURE P12

13. [2.21] The voltage across the 22.5 Ω resistor in the circuit in Fig. P13 is 90 V, positive at the upper terminal.

a) Find the power dissipated in each resistor.

b) Find the power supplied by the 240 V ideal voltage source.

c) Verify that the power supplied equals the total power dissipated.

FIGURE P13

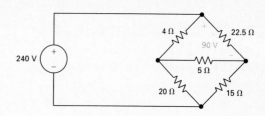

14. [2.26] Find v_1 and v_g in the circuit shown in Fig. P14 when v_0 equals 250 mV.

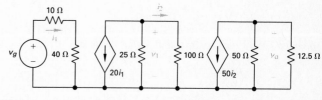

FIGURE P14

15. [2.28] For the circuit shown in Fig. P15, (a) calculate i_Δ and v_o and (b) show that the power developed equals the power absorbed.

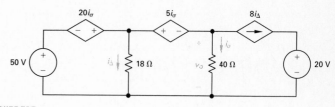

FIGURE P15

16. [2.31] A person accidently grabs conductors connected to each end of a dc voltage source, one in each hand.

a) Using the resistance values for the human body provided in Chapter 2 of the text (Fig. 2.26), what is the minimum source voltage that can produce electrical shock sufficient to cause paralysis, preventing the person from letting go of the conductors?

b) Is there a significant risk of this type of accident occurring while servicing a personal computer, which typically has 5 V and 12 V sources?

17. [3.4]

  a) In the circuit in Fig. P17 find the equivalent resistance R_{ab}.

  b) Find the power delivered by the source.

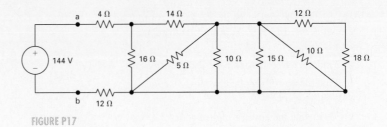

FIGURE P17

18. [3.7] Find the equivalent resistance R_{ab} for the circuit in Fig. P18.

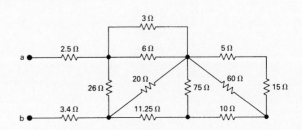

FIGURE P18

19. [3.10] Find v_o and v_g in the circuit in Fig. P19.

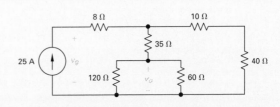

FIGURE P19

20. [3.17] The no-load voltage in the voltage-divider circuit shown in Fig. P20 is 20 V. The smallest load resistor that is ever connected to the divider is 48 kΩ. When the divider is loaded, v_o is not to drop below 16 V.

a) Design the divider circuit to meet the specifications just mentioned. Specify the numerical values of R_1 and R_2.

b) Assume that the power ratings of commercially available resistors are 1/16, 1/8, 1/4, 1, and 2 W. Design the numerical values of R_1 and R_2 and their power ratings to meet the specifications enumerated.

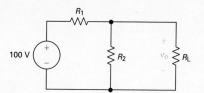

FIGURE P20

21. [3.30] The ammeter in the circuit in Fig. P21 has a resistance of 0.5 Ω. What is the percentage of error in the reading of this ammeter if

$$\% \text{ error} = \left(\frac{\text{measured value}}{\text{true value}} - 1 \right) \times 100?$$

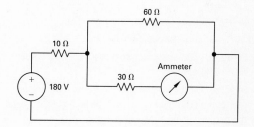

FIGURE P21

22. [3.44] Find the power dissipated in the 9 Ω resistor in the circuit in Fig. P22.

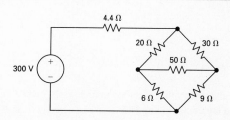

FIGURE P22

23. [3.50] Find i_o and the power dissipated in the 30 Ω resistor in the circuit in Fig. P23.

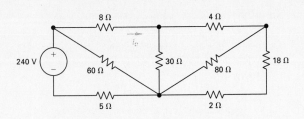

FIGURE P23

24. [3.61] The design equations for the bridged-tee attenuator circuit in Fig. P24 are

$$R_{ab} = R_L$$

when

$$R_2 = \frac{2RR_L^2}{3R^2 - R_L^2} \quad \text{and} \quad \frac{v_o}{v_i} = \frac{3R - R_L}{3r + R_L}$$

when R_2 satisfies the equation above.

a) Design a fixed attenuator so that $v_i = 3.5v_o$ when $R_L = 300\ \Omega$.

b) Assume that the voltage applied to the input of the pad designed in part (a) is 42 V. Which resistor in the pad dissipates the most power?

c) How much power is dissipated in the resistor in part (b)?

d) Which resistor in the pad dissipates the least power?

e) How much power is dissipated in the resistor in part (d)?

FIGURE P24

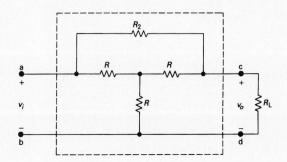

25. [4.6]

a) Use the node-voltage method to find v_1, v_2, and v_3 in the circuit in Fig. P25.

b) How much power does the 640 V voltage source deliver to the circuit?

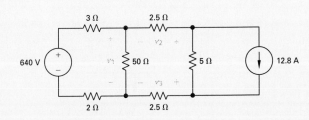

FIGURE P25

26. [4.18]

   a) Use the node-voltage method to find v_o in the circuit in Fig. P26.

   b) Find the power absorbed by the dependent source.

   c) Find the total power developed by the independent sources.

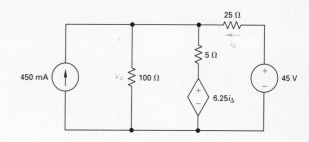

FIGURE P26

27. [4.29]

   a) Use the mesh-current method to find how much power the 12 A current source delivers to the circuit in Fig. P27.

   b) Find the total power delivered to the circuit.

   c) Check your calculations by showing that the total power developed in the circuit equals the total power dissipated.

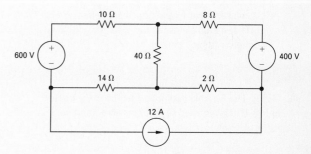

FIGURE P27

28. [4.42] Use the mesh-current method to find the power developed in the dependent-voltage source in the circuit in Fig. P28.

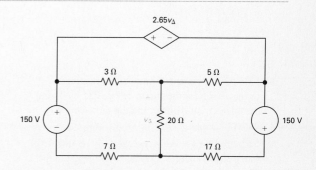

FIGURE P28

29. [4.47] The circuit in Fig. P29 is a direct-current version of a typical three-wire distribution system. The resistors R_a, R_b, and R_c represent the resistances of the three conductors that connect the three loads R_1, R_2, and R_3 to the 125/250 V voltage supply. The resistors R_1 and R_2 represent loads connected to the 125 V circuits, and R_3 represents a load connected to the 250 V circuit.

a) Calculate v_1, v_2, and v_3.

b) Calculate the power delivered to R_1, R_2, and R_3.

c) What percentage of the total power developed by the sources is delivered to the loads?

d) The R_b branch represents the neutral conductor in the distribution circuit. What adverse effect occurs if the neutral conductor is opened? (*Hint:* Calculate v_1 and v_2 and note that appliances or loads designed for use in this circuit would have a nominal voltage rating of 125 V.)

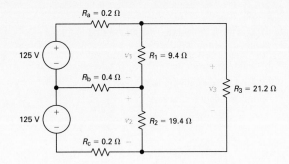

FIGURE P29

30. [4.50] The variable dc-voltage source in the circuit in Fig. P30 is adjusted so that i_o is zero. Find the value of V_{dc}.

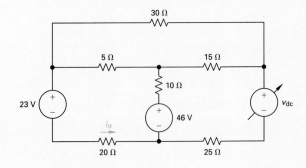

FIGURE P30

31. [4.58]

a) Find the Thévenin equivalent with respect to terminals a,b for the circuit in Fig. P31 by finding the open-circuit voltage and the short-circuit current.

b) Solve for the Thévenin resistance by removing the independent sources. Compare your result to the Thévenin resistance found in part (a).

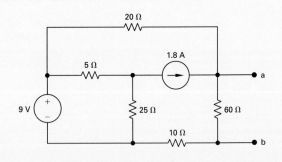

FIGURE P31

32. [4.60] Determine i_o and v_o in the circuit in Fig. P32 when R_o is 0, 2, 6, 10, 15, 20, 30, 40, 50, and 70 Ω.

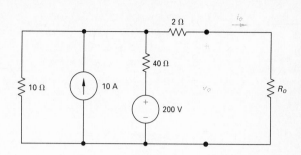

FIGURE P32

33. [4.69] You can also determine a Thévenin equivalent from measurements made at the pair of terminals of interest. Assume that the following measurements were made at terminals a,b in the circuit in Fig. P33.

When a 20 kΩ resistor is connected to terminals a,b, the voltage v_{ab} is measured and found to be 20 V.

When a 7.5 kΩ resistor is connected to terminals a,b, the voltage is measured and found to be 15 V.

Find the Thévenin equivalent of the network with respect to terminals a,b.

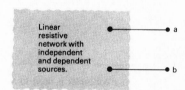

FIGURE P33

34. [4.73] The variable resistor in the circuit in Fig. P34 is adjusted for maximum power transfer to R_L.

a) Find the value of R_L.

b) Find the maximum power that can be delivered to R_L.

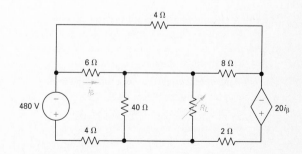

FIGURE P34

35. [4.96] Find the power absorbed by the 5-A current source in the circuit in Fig. P35.

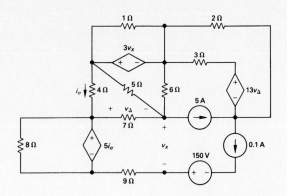

FIGURE P35

36. [5.1] The operational amplifier in the circuit in Fig. P36 is ideal.

a) Calculate v_o if $v_a = 1.0$ V and $v_b = 0$ V.

b) Calculate v_o if $v_a = 2.0$ V and $v_b = 0$ V.

c) Calculate v_o if $v_a = 2.5$ V and $v_b = 3$ V.

d) Calculate v_o if $v_a = 3.0$ V and $v_b = 2$ V.

e) Calculate v_o if $v_a = 1.5$ V and $v_b = 2.5$ V.

f) If $v_b = 4.0$ V, specify the range of v_a such that the amplifier does not saturate.

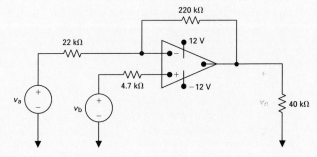

FIGURE P36

37. [5.5] A circuit designer claims that the circuit in Fig. P37 will produce an output voltage that will vary between ±9 as v_g varies between 0 and 6 V. Assume that the operational amplifier is ideal.

a) Plot the output voltage v_o as a function of the input voltage v_g for $0 \leq v_g \leq 6$ V.

b) Do you agree with the designer's claim?

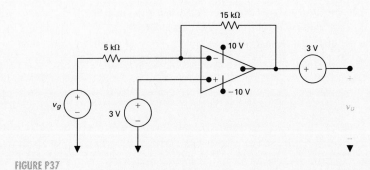

FIGURE P37

38. [5.10] The operational amplifier in Fig. P38 is ideal.

  a) Find v_o if $v_a = 1.2$ V, $v_b = -1.5$ V, and $v_c = 4$ V.

  b) The voltages v_a and v_c remain at 1.2 V and 4 V respectively. What are the limits on v_b if the operational amplifier operates within its linear region?

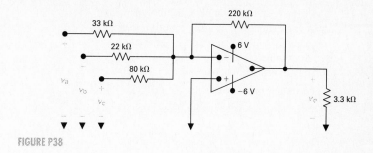

FIGURE P38

39. [5.15] The variable resistor R_o in the circuit in Fig. P39 is adjusted until the source current i_g is zero. The operational amplifiers are ideal and $0 \le v_g \le 1.2$ V.

  a) What is the value of R_o?

  b) If $v_g = 1.0$ V, how much power (in μW) is dissipated in R_o?

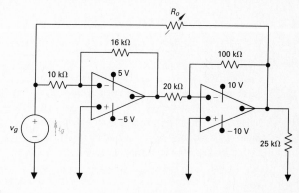

FIGURE P39

40. [5.26] The operational amplifier in the circuit in Fig. P40 is ideal. Plot v_o versus α when $R_f = 4\,R_1$ and $v_g = 10$ V. Use increments of 0.1 and note by hypothesis that $0 \le \alpha \le 1.0$.

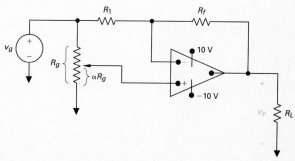

FIGURE P40

41. [5.34] The operational amplifier in the noninverting amplifier circuit of Fig. P41 has an input resistance of 560 kΩ, an output resistance of 8 kΩ, and an open-loop gain of 50,000. Assume that the op amp is operating in its linear region. Calculate the following:

a) The voltage gain (v_o/v_g);

b) The inverting and noninverting input voltages v_1 and v_2 (in mV) if $v_g = 1$ V;

c) The difference $(v_2 - v_1)$ in microvolts when $v_g = 1$ V;

d) The current drain in picoamperes on the signal source v_g when $v_g = 1$ V.

Repeat parts (a) through (d) assuming an ideal op amp.

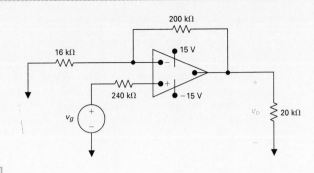

FIGURE P41

42. [5.36]

a) Find the Thévenin equivalent circuit with respect to the output terminals a,b for the inverting amplifier in Fig. P42. The dc signal source has a value of 200 mV. The operational amplifier has an input resistance of 400 kΩ, an output resistance of 800 Ω, and an open loop gain of 10,000.

b) What is the output resistance of the inverting amplifier?

c) What is the resistance (in ohms) seen by the signal source v_g when the load at terminals a,b is 220 Ω?

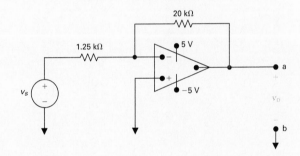

FIGURE P42

43. [5.41] Find v_o and i_o in the circuit in Fig. P43 if the operational amplifiers are ideal.

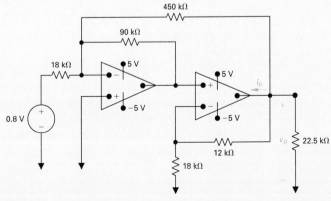

FIGURE P43

44. [5.43] The voltage v_g shown in Fig. P44(a) is applied to the inverting amplifier shown in Fig. P44(b). Plot v_o versus t, assuming that the operational amplifier is ideal.

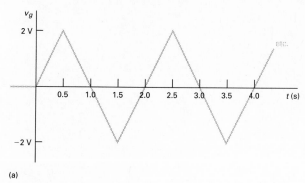

(a)

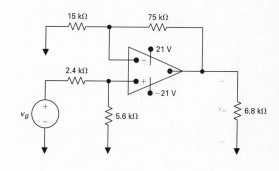

(b)

FIGURE P44

45. [5.44] The signal voltage v_g in the circuit in Fig. P45 is described by the following equations:

$$v_g = 0, \quad t \le 0;$$

$$v_g = 10 \sin(\pi/3)t \text{ V}, \quad 0 \le t \le \infty.$$

Plot v_o versus t, assuming that the operational amplifier is ideal.

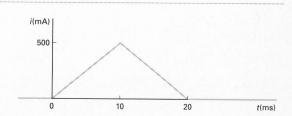

FIGURE P45

46. [6.2] The triangular pulse shown in Fig. P46 is applied to a 40 mH inductor.

a) Find the expressions that describe $i(t)$ in the four intervals $t < 0, 0 \le t \le 10$ ms, 10 ms $\le t \le 20$ ms, and $t > 20$ ms.

b) Find the expressions for the inductor voltage, power, and energy. Use the passive sign convention.

c) Plot the inductor current, voltage, power, and energy.

FIGURE P46

47. [6.6] The current in a 20 mH inductor is known to be
−10 A for $t \leq 0$ and
$i = 7 + (15 \sin 140t - 35 \cos 140t)e^{-20t}$ mA for
$t \geq 0$. Assume the passive sign convention.

a) At what instant of time is the voltage across the
    inductor maximum?

b) What is the maximum voltage?

48. [6.8] The current in and the voltage across a 2.5 mH
inductor are known to be zero for $t \leq 0$. The voltage
across the inductor is shown in the graph in Fig. P48
for $t \geq 0$.

a) Find the expressions for the current as a function
   of time in the intervals $0 \leq t \leq 2$ s, 2 s $\leq t \leq 6$ s,
   6 s $\leq t \leq 10$ s, 10 s $\leq t \leq 12$ s, and 12 s
   $\leq t \leq \infty$.

b) For $t > 0$, what is the current in the inductor when
   the voltage is zero?

c) Plot i versus t for $0 \leq t \leq \infty$.

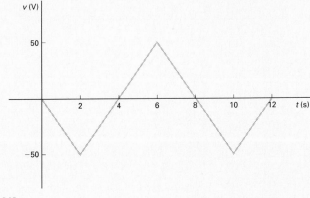

FIGURE P48

49. [6.15] The rectangular-shaped current pulse shown
in Fig. P49 is applied to a 0.2 μF capacitor. The
initial voltage on the capacitor is a 40 V drop in the
reference direction of the current. Assume the passive
sign convention. Find the expression for the capacitor
voltage for each of the following time intervals:

a) $0 \leq t \leq 100$ μs;

b) 100 μs $\leq t \leq 300$ μs;

c) 300 μs $\leq t \leq \infty$.

Plot $v(t)$ over the interval -100 μs $\leq t \leq 500$ μs.

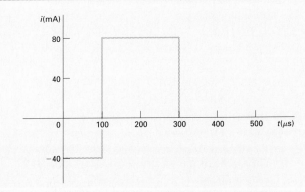

FIGURE P49

50. [6.17] The initial voltage on the 0.5 μF capacitor in
Fig. P50 (a) is 20 V. The capacitor current has the
waveform shown in Fig. P50 (b).

a) How much energy, in microjoules, is stored in the
capacitor at $t = 500 \ \mu$s?

b) How much energy, in microjoules, is stored in the
capacitor at $t = \infty$?

(a)

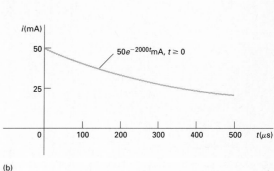

(b)

FIGURE P50

51. [6.20] Assume that the initial energy stored in the
inductors in Fig. P51 is zero. Find the equivalent
inductance with respect to terminals a,b.

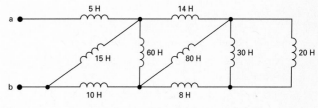

FIGURE P51

52. [6.26] Find the equivalent capacitance with respect
to terminals a,b for the circuit in Fig. P52.

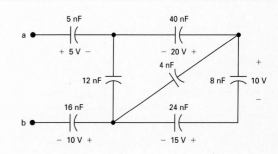

FIGURE P52

53. [6.31] The current in the circuit in Fig. P53 is known to be

$$i_o = 50e^{-8000t}[\cos 6000t + 2\sin 6000t] \text{ A}$$

for $t \geq 0^+$. Find $v_1(0^+)$ and $v_2(0^+)$.

FIGURE P53

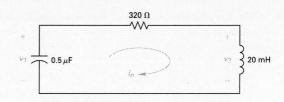

54. [7.3] The switch in the circuit in Fig. P54 has been open for a long time. At $t = 0$ the switch is closed.

a) Determine $i_o(0^+)$ and $i_o(\infty)$.

b) Determine $i_o(t)$ for $t \geq 0^+$.

c) How many microseconds after the switch has been closed will the current in the switch equal 3.2 A?

FIGURE P54

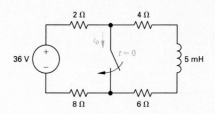

55. [7.9] In the circuit in Fig. P55, the switch has been closed for a long time before opening at $t = 0$.

a) Find the value of L so that $v_o(t)$ equals $0.2v_o(0^+)$ when $t = 2$ ms.

b) Find the percentage of the stored energy that has been dissipated in the 20 Ω resistor when $t = 2$ ms.

FIGURE P55

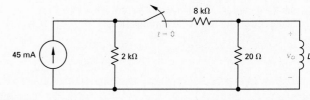

56. [7.19] The 240 V, 2 Ω source in the circuit in Fig. P56 is inadvertently short-circuited at its terminals a,b. At the time the fault occurs, the circuit has been in operation for a long time.

a) What is the initial value of the current i_{ab} in the short-circuit connection between terminals a,b?

b) What is the final value of the current i_{ab}?

c) How many microseconds after the short-circuit has occurred is the current in the short equal to 114 A?

FIGURE P56

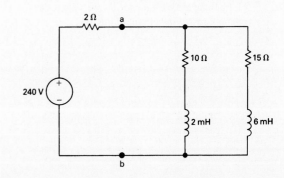

57. [7.20] In the circuit in Fig. P57 the voltage and current expressions are

$$v = 100e^{-1000t} \text{ V}, \quad t \geq 0;$$
$$i = 5e^{-1000t} \text{ mA}, \quad t \geq 0^+.$$

Find (a) R, (b) C, (c) τ (in milliseconds), (d) the initial energy stored in the capacitor, and (e) how many milliseconds it takes to dissipate 80% of the initial energy stored in the capacitor.

58. [7.31] At the time the switch is closed in the circuit in Fig. P58, the voltage across the paralleled capacitors is 50 V and the voltage on the 0.25μF capacitor is 40 V.

a) What percentage of the initial energy stored in the three capacitors is dissipated in the 24 kΩ resistor?

b) Repeat part (a) for the 0.4 and the 16 kΩ resistors.

c) What percentage of the initial energy is trapped in the capacitors?

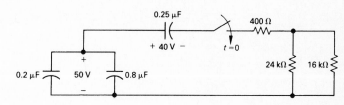

FIGURE P58

59. [7.44] There is no energy stored in the inductors L_1 and L_2 at the time the switch is opened in the circuit in Fig. P59.

a) Find the expressions for the currents $i_1(t)$ and $i_2(t)$ for $t \geq 0$.

b) Find the values $i_1(\infty)$ and $i_2(\infty)$

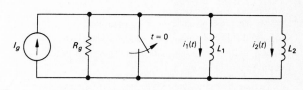

FIGURE P59

60. [7.65] There is no energy stored in the capacitor in the circuit in Fig. P60 when switch 1 closes at $t = 0$. Ten microseconds later switch 2 closes. Find $v_o(t)$ for $t \geq 0$.

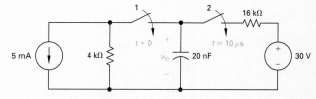

FIGURE P60

61. [7.69] The voltage waveform in Fig. P61(a) is applied to the circuit of Fig. P61(b). The initial voltage on the capacitor is zero.

a) Find $v_o(t)$.

b) Plot $v_o(t)$ versus t to verify your solution in (a).

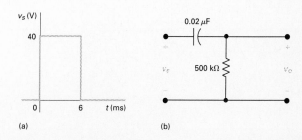

(a)   (b)

FIGURE P61

62. [7.87] The voltage pulse in Fig. P62(a) is applied to the ideal integrating amplifier in Fig. P62(b). Derive the numerical expressions for $v_o(t)$ for the time intervals (a) $t < 0$; (b) $0 \le t \le 250$ ms; (c) 250 ms $\le t \le 500$ ms; and (d) 500 ms $\le t \le \infty$ when $v_o(0) = 0$.

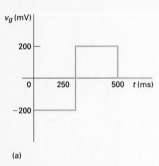

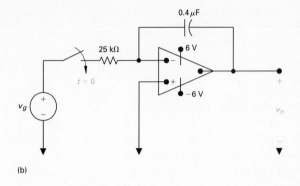

(a)   (b)

FIGURE P62

63. [7.89] The voltage source in the circuit in Fig. P63(a) is generating the triangular waveform shown in Fig. P63(b). Assume that the energy stored in the capacitor is zero at $t = 0$.

a) Derive the numerical expressions for $v_o(t)$ for the following time intervals: $0 \le t \le 1$ μs; 1 μs $\le t \le 3$ μs; and 3 μs $\le t \le 4$ μs.

b) Plot the output waveform between 0 and 4 μs.

c) If the triangular input voltage continues to repeat itself for $t > 4$ μs, what would the output voltage look like?

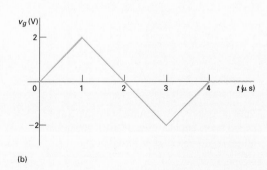

(a)

FIGURE P63

(b)

64. [8.22-24] Assume that at the instant the 60 mA dc current source is applied to the circuit in Fig. P64, the initial current in the 50 mH inductor is −45 mA and the initial voltage on the capacitor is 15 V, positive at the upper terminal.

a) Find the expression for $i_L(t)$ for $t \geq 0$ if $R = 200 \ \Omega$.

b) Repeat part (a) if R is increased to 312.5 Ω.

c) Repeat part (a) if R is increased to 250 Ω.

FIGURE P64

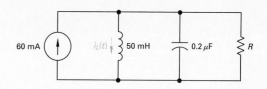

65. [8.36] In the circuit in Fig. P65, the resistor is adjusted for critical damping. The initial capacitor voltage is 15 V, and the initial inductor current is 6 mA.

a) Find the numerical value of R.

b) Find the numerical values of i and di/dt immediately after the switch is closed.

c) Find $v_C(t)$ for $t \geq 0$.

FIGURE P65

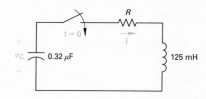

66. [8.49] The voltage signal in Fig. P66(a) is applied to the cascaded integrating amplifiers in Fig. P66(b). There is no energy stored in the capacitors at the instant the signal is applied.

a) Find the numerical expressions for $v_o(t)$ and $v_{o1}(t)$ for the time intervals $0 \leq t \leq 0.5$ s and 0.5 s $\leq t \leq t_{\text{sat}}$.

b) Compute the value of t_{sat}.

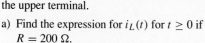

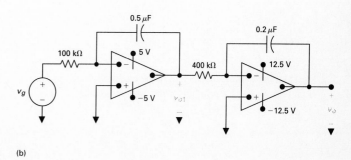

(a)

(b)

FIGURE P66

67. [8.51] Find the differential equation that relates the output voltage to the input voltage for the circuit in Fig. P67. Compare the result with the two-stage amplifier in Problem 66.

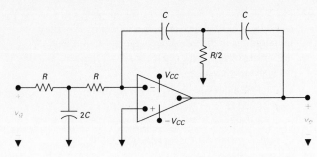

FIGURE P67

68. [9.2] On a single graph, plot $v = 60\cos(\omega t + \phi)$ versus ωt for $\phi = -60°, -30°, 0°, +30°$, and $60°$. Comment on the direction in which the voltage function shifts as ϕ becomes more positive.

69. [9.9] A 400 Hz sinusoidal voltage with a maximum amplitude of 100 V is applied across the terminals of an inductor. The maximum amplitude of the steady-state current in the inductor is 20 A.

a) What is the frequency of the inductor current?

b) What is the phase angle of the voltage?

c) What is the phase angle of the current?

d) What is the inductive reactance of the inductor?

e) What is the inductance of the inductor in millihenrys?

f) What is the impedance of the inductor?

70. [9.13] Find the steady-state expression for $i_o(t)$ in the circuit in Fig. P70 if $v_s = 500\sin 4000t$ mV. Verify your result by plotting $i_o(t)$ versus t.

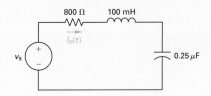

FIGURE P70

71. [9.15] Find the impedance Z_{ab} in the circuit seen in Fig. P71. Express Z_{ab} in both polar and rectangular form.

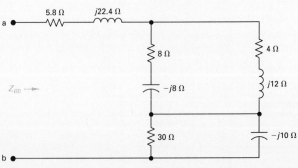

FIGURE P71

72. [9.20] The circuit in Fig. P72 is operating in the sinusoidal steady state. The capacitor is adjusted until the current i_g is in phase with the sinusoidal voltage v_g.

a) Specify the values of capacitance in microfarads if $v_g = 80 \cos 5000t$ V.

b) Find the steady-state expressions for i_g when C has the values found in part (a).

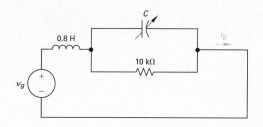

FIGURE P72

73. [9.24]

a) The frequency of the source voltage in the circuit in Fig. P73 is adjusted until i_g is in phase with v_g. What is the value of ω in rad/s?

b) If $v_g = 20 \cos \omega t$ V, where ω is the frequency found in part (a), what is the steady-state expression for v_o?

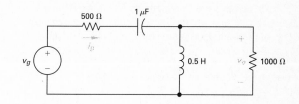

FIGURE P73

74. [9.32] The phasor current $\mathbf{I}_a$ in the circuit in Fig. P74 is $2 \underline{/-0°}$ A.

a) Find $\mathbf{I}_b$, $\mathbf{I}_c$, and $\mathbf{V}_g$.

b) If $\omega = 800$ rad/s, find the expressions for $i_b(t)$, $i_c(t)$, and $v_g(t)$.

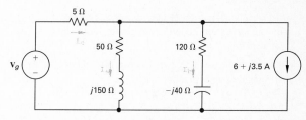

FIGURE P74

75. [9.45] Find the branch currents $\mathbf{I}_a$, $\mathbf{I}_b$, and $\mathbf{I}_c$ in the circuit in Fig. P75.

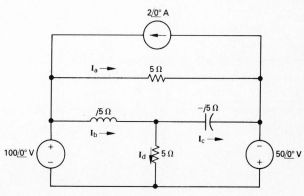

FIGURE P75

76. [9.51] Find the Thévenin equivalent circuit with respect to terminals a,b for the circuit in Fig. P76.

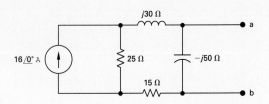

FIGURE P76

77. [9.62] Create a phasor diagram to show what happens to the magnitude and phase angle of the voltage v_o in the circuit in Fig. P77 as R_x is varied from zero to infinity. The amplitude and phase angle of the source voltage are held constant as R_x varies.

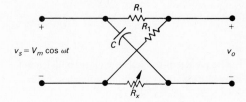

FIGURE P77

78. [9.65] The operational amplifier in the circuit in Fig. P78 is ideal. The voltage of the ideal sinusoidal source is $v_g = 30 \cos 10^6 t$ V.

a) How small can C_o be before the steady-state output voltage no longer has a pure sinusoidal waveform?

b) For the value of C_o found in part (a), find the steady-state expression for v_o.

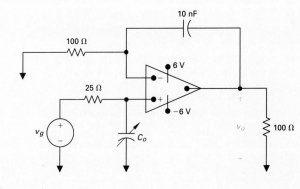

FIGURE P78

79. [9.68]

a) Find the input impedance Z_{ab} for the circuit in Fig. P79.

b) If Z is a pure capacitive element, what is the capacitance looking into terminals a,b?

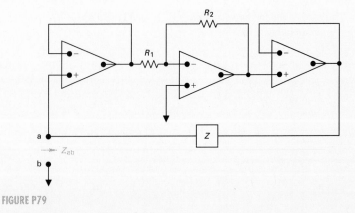

FIGURE P79

80. [10.3] A load consisting of a 1.25 kΩ resistor in parallel with a 405 mH inductor is connected across the terminals of a sinusoidal voltage source v_g, where $v_g = 90 \cos 2500t$ V.

a) What is the peak value of the instantaneous power delivered by the source?

b) What is the peak value of the instantaneous power absorbed by the source?

c) What is the average power delivered to the load?

d) What is the reactive power?

e) Does the load absorb or generate magnetizing vars?

f) What is the power factor of the load?

g) What is the reactive factor of the load?

81. [10.5] Find the rms value of the periodic current in Fig. P81.

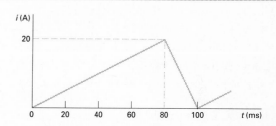

FIGURE P81

82. [10.12] Find the average power delivered by the ideal current source in the circuit in Fig. P82 if $i_g = 4 \cos 5000t$ mA.

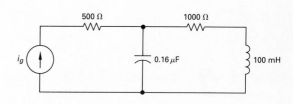

FIGURE P82

83. [10.17] Three loads are connected in parallel across a 250 V (rms) line, as shown in Fig. P83. Load 1 absorbs 3 kW and 4 kVAR. Load 2 absorbs 2.5 kVA at 0.6 pf lead. Load 3 absorbs 1.5 kW at unity power factor.

a) Find the impedance that is equivalent to the three parallel loads.

b) Find the power factor of the equivalent load from the line's input terminals.

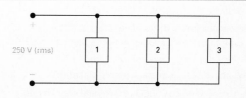

FIGURE P83

84. [10.21] The three parallel loads in the circuit in Fig. P84 can be described as follows: Load 1 is absorbing an average power of 7.5 kW and 9 kVAR of magnetizing vars; load 2 is absorbing an average power of 2.1 kW and is generating 1.8 kVAR of magnetizing reactive power; load 3 consists of a 48 Ω resistor in parallel with an inductive reactance of 19.2 Ω. Find the rms magnitude and the phase angle of $\mathbf{V}_g$ if $\mathbf{V}_o = 480\underline{/-0°}$ V (rms).

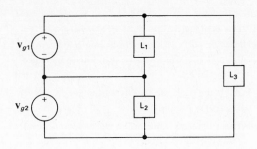

FIGURE P84

85. [10.27]

a) Find the average power dissipated in the line in the circuit in Fig. P85.

b) Find the capacitive reactance that when connected in parallel with the load will make the load look purely resistive.

c) What is the equivalent impedance of the load in part (b)?

d) Find the average power dissipated in the line when the capacitive reactance is connected across the load.

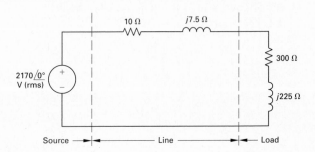

FIGURE P85

86. [10.30] The peak amplitude of the sinusoidal voltage source in the circuit shown in Fig. P86 is $100\sqrt{2}$ V and its period is 250π μs. The load resistor can be varied from 0 to 200 Ω, and the load capacitor can be varied from 1 to 4 μF.

a) Calculate the average power delivered to the load when $R_L = 100\,\Omega$ and $C_L = 4\,\mu$F.

b) Determine the settings of R_L and C_L that will result in the most average power being transferred to R_L.

c) What is the most average power in part (b)?

d) If there are no constraints on R_L and C_L, what is the maximum average power that can be delivered to a load?

e) Calculate the values of R_L and C_L for the condition of part (d).

f) Is the average power calculated in part (d) larger than that calculated in part (c)?

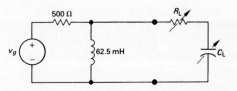

FIGURE P86

87. [10.35] The load impedance Z_L for the circuit in
Fig. P87 is adjusted until maximum average power is
delivered to Z_L.

a) Find the maximum average power delivered to
Z_L.

b) What percentage of the total power developed in
the circuit is delivered to Z_L?

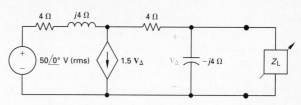

FIGURE P87

88. [10.37] A factory has an electrical load of 1600 kW
at a lagging power factor of 0.8. An additional
variable-power factor load is to be added to the
factory. The new load will add 320 kW to the real
power load of the factory. The power factor of the
added load is to be adjusted so that the overall power
factor of the factory is 0.96 leading.

a) Specify the reactive power associated with the
added load.

b) Does the added load absorb or deliver
magnetizing vars?

89. [11.6]

a) Find $\mathbf{I}_o$ in the circuit in Fig. P89.

b) Find $\mathbf{V}_{BN}$.

c) Find $\mathbf{V}_{BC}$.

d) Is the circuit a balanced or unbalanced three-phase
system?

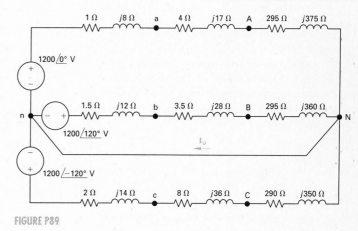

FIGURE P89

90. [11.11] A balanced Y-connected load having an
impedance of $72 + j21\ \Omega/\phi$ is connected in parallel
with a balanced Δ-connected load having an
impedance of $150\ \underline{/-0°}\ \Omega/\phi$. The paralleled loads
are fed from a line having an impedance of $j1\ \Omega/\phi$.
The magnitude of the line-to-neutral voltage of the
Y load is 7650 V.

a) Calculate the magnitude of the current in the line
feeding the loads.

b) Calculate the magnitude of the phase current in
the Δ-connected load.

c) Calculate the magnitude of the phase current in
the Y-connected load.

d) Calculate the magnitude of the line voltage at the
sending end of the line.

91. [11.18] A three-phase line has an impedance of $5 + j4\,\Omega/\phi$. The line feeds two balanced three-phase loads that are connected in parallel. The first load is absorbing a total of 691.20 kW and delivering 201.6 kVAR magnetizing vars. The second load is Δ-connected and has an impedance of $622.06 + j181.44\,\Omega/\phi$. The line-to-neutral voltage at the load end of the line is 7200 V. What is the magnitude of the line voltage at the source end of the line?

92. [11.23] The three pieces of computer equipment described below are installed as part of a computation center. Each piece of equipment is a balanced three-phase load rated at 208 V. Calculate (a) the magnitude of the line current supplying these three devices and (b) the power factor of the combined load.

Disk: 4.864 kW at 0.79 pf lag;
Drum: 17.636 kVA at 0.96 pf lag;
CPU: line current 73.8 A, 13.853 kVAR.

93. [11.26] The total power delivered to a balanced three-phase load when operating at a line voltage of $4800\sqrt{3}$ V is 900 kW at a lagging power factor of 0.60. The impedance of the distribution line supplying the load is $0.6 + j4.8\,\Omega/\phi$. Under these operating conditions, the drop in the magnitude of the line voltage between the sending end and the load end of the line is excessive. To compensate for the excessive voltage drop, a bank of Δ-connected capacitors is placed in parallel with the load. The capacitor bank is designed to furnish 1200 kVAR of magnetizing reactive power when operated at a line voltage of $4800\sqrt{3}$ V.

a) What is the magnitude of the voltage at the sending end of the line when the load is operating at a line voltage of $4800\sqrt{3}$ V and the capacitor bank is disconnected?

b) Repeat part (a), with the capacitor bank connected.

c) What is the average power efficiency of the line in part (a)?

d) What is the average power efficiency in part (b)?

e) If the system is operating at a frequency of 60 Hz, what is the size of each capacitor in microfarads?

94. [11.32]

a) Find the reading of each wattmeter in the circuit shown in Fig. 94 if $Z_A = 20\underline{/30°}\,\Omega$, $Z_B = 60\underline{/0°}\,\Omega$, and $Z_C = 40\underline{/-30°}\,\Omega$.

b) Show that the sum of the wattmeter readings equals the total average power delivered to the unbalanced three-phase load.

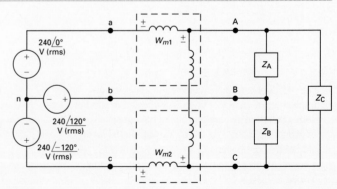

FIGURE P94

95. [12.8] A series combination of a 60 Ω resistor and a 15.625 nF capacitor is connected to a sinusoidal voltage source by a linear transformer. The source is operating at a frequency of 800 krad/s. At this frequency the internal impedance of the source is $10 + j27.2\ \Omega$. The rms voltage at the terminals of the source is 80 V when it is not loaded. The parameters of the linear transformer are: $R_1 = 15.6\ \Omega$, $L_1 = 90\ \mu H$, $R_2 = 30\ \Omega$, $L_2 = 250\ \mu H$, and $M = 75\ \mu H$.

a) What is the value of the impedance reflected into the primary?

b) What is the value of the impedance seen from the terminals of the practical source?

c) What is the rms magnitude of the voltage across the load impedance?

d) What percentage of the average power developed by the practical source is delivered to the load impedance?

96. [12.9]

a) Find the steady-state expressions for the currents i_g and i_L in the circuit in Fig. P96 when $v_g = 200 \cos 10,000t$ V.

b) Find the coefficient of coupling.

c) Find the energy stored in the magnetically coupled coils at $t = 50\pi\ \mu s$ and $t = 100\pi\ \mu s$.

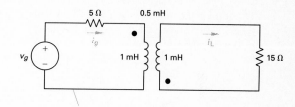

FIGURE P96

97. [12.14]

a) For the circuit in Fig. P97, find the Thévenin equivalent with respect to terminals c,d.

b) Find the average power developed by the sinusoidal voltage source if an impedance equal to the conjugate of the Thévenin impedance is connected to terminals c,d.

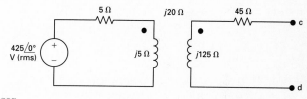

FIGURE P97

98. [12.15] The impedance Z_L in the circuit in Fig. P98 is adjusted for maximum average power transfer to Z_L. The internal impedance of the sinusoidal voltage source is $20 + j35\ \Omega$.

a) What is the maximum average power delivered to Z_L?

b) What percentage of the average power delivered to the linear transformer is delivered to Z_L?

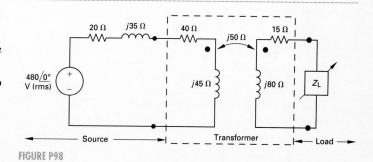

FIGURE P98

99. [12.23] Find the impedance seen by the ideal voltage
source in the circuit in Fig. P99 when Z_o is adjusted
for maximum average power transfer to Z_o.

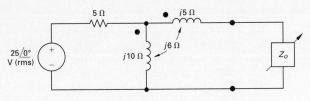

FIGURE P99

100. [12.28] Find the average power dissipated in the
1 Ω resistor in the circuit in Fig. P100. Check your
answer by showing that the total power developed
equals the total power absorbed.

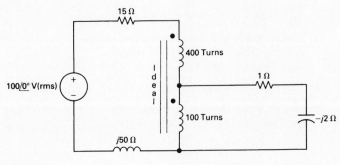

FIGURE P100

101. [12.34] The sinusoidal current source in the circuit
in Fig. P101 is operating at a frequency of 20 krad/s.
The variable capacitor in the circuit is adjusted until
the average power delivered to the 100 Ω resistor is
as large as possible.

a) Find the value of C in μF.

b) When C has the value found in part (a), what is
the average power delivered to the 100 Ω
resistor?

c) Replace the 100 Ω resistor with a variable
resistor R_o. Specify the value of R_o so that
maximum average power is delivered to R_o.

d) What is the maximum average power that can be
delivered to R_o?

FIGURE P101

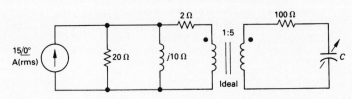

102. [13.3] You can use step functions to define a "window" function. Thus $u(t-1) - u(t-4)$ defines a window one unit high and three units wide located on the time axis between 1 and 4.
A function $f(t)$ is defined as follows:

$$f(t) = 0, t \leq 0;$$
$$f(t) = -20t, 0 \leq t \leq 1 \text{ s};$$
$$f(t) = 20\cos \pi t, 1 \text{ s} \leq t \leq 4 \text{ s};$$
$$f(t) = 100 - 20t, 4 \text{ s} \leq t \leq 5 \text{ s};$$
$$f(t) = 0, 5 \text{ s} \leq t \leq \infty.$$

Plot $f(t)$ over the interval $-1 \text{ s} \leq t \leq 6 \text{ s}$.

103. [13.17]

a) Find the Laplace transform of the function illustrated in Fig. P103.

b) Find the Laplace transform of the first derivative of the function illustrated in Fig. P103.

c) Find the Laplace transform of the second derivative of the function illustrated in Fig. P103.

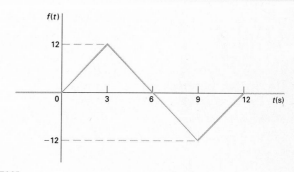

FIGURE P103

104. [13.27] Find $f(t)$ for the following functions:

a) $F(s) = \dfrac{50(s+5)}{s(s+1)^2}$

b) $F(s) = \dfrac{5(s+2)^2}{s(s+1)^3}$

c) $F(s) = \dfrac{400}{s(s^2+4s+5)^2}$

105. [14.7] Find the poles and zeros of the impedance seen looking into terminals a,b of the circuit shown in Fig. P105.

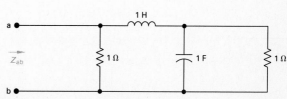

FIGURE P105

106. [14.16] There is no energy stored in the circuit in Fig. P106 at the time the voltage source is energized.

  a) Find I_o and V_o.

  b) Plot i_o and v_o for $t \geq 0$.

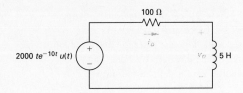

FIGURE P106

107. [14.21] The switch in the circuit in Fig. P107 has been closed for a long time before opening at $t = 0$. Find v_o for $t \geq 0$.

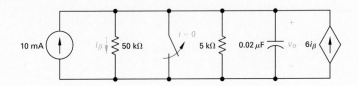

FIGURE P107

108. [14.26] There is no energy stored in the circuit in Fig. P108 at the time the sources are energized.

  a) Find $I_1(s)$ and $I_2(s)$.

  b) Plot $i_1(t)$ and $i_2(t)$ for $t \geq 0$.

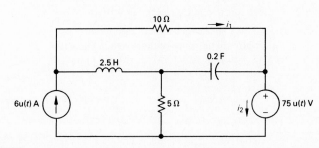

FIGURE P108

109. [14.36] The switch in the circuit in Fig. P109 has been closed for a long time before opening at $t = 0$. Find and plot v_o.

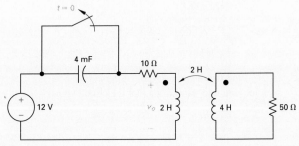

FIGURE P109

110. [14.41] The operational amplifier in the circuit in Fig. P110 is ideal. There is no energy stored in the circuit at the time it is energized. If $v_g = 16000tu(t)$ V, find V_o. Find and plot v_o and determine how long it takes to saturate the operational amplifier, and how small the rate of increase in v_g must be to prevent saturation.

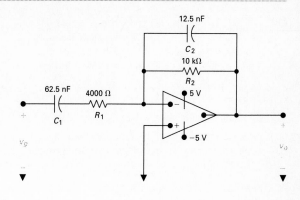

FIGURE P110

111. [14.44] The operational amplifier in the circuit in Fig. P111 is ideal. There is no energy stored in the capacitors at the instant the circuit is energized.

a) Find and plot v_o if $v_{g1} = 16u(t)$ V and $v_{g2} = 8u(t)$ V.

b) How many milliseconds after the two voltage sources are turned on does the op amp saturate?

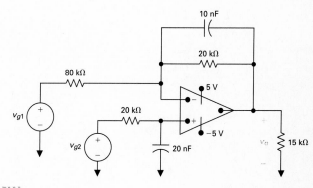

FIGURE P111

112. [14.55]

a) Find the transfer function I_o/I_g as a function of μ for the circuit shown in Fig. P112.

b) Find the largest value of μ that will produce a bounded output signal for a bounded input signal.

c) Find and plot i_o for $\mu = -3, 0, 4, 5$, and 6 if $i_g = 5u(t)$ A.

FIGURE P112

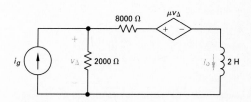

113. [14.58] Assume that the voltage impulse response
of a circuit can be modeled by the triangular
waveform shown in Fig. P113. The voltage input
signal to this circuit is the step function $10u(t)$ V.

   a) Use the convolution integral to determine the
expressions for the output voltage.

   b) Plot the output voltage over the interval 0 to 15 s.

   c) Repeat parts (a) and (b) if the area under the
voltage impulse response stays the same but the
width of the impulse response narrows to 4 s.

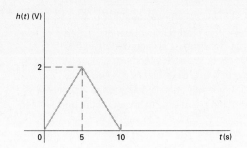

FIGURE P113

114. [14.65] The current source in the circuit in
Fig. P114(a) is generating the waveform in
Fig. P114(b). Use convolution to find v_o at $t = 5$
ms.

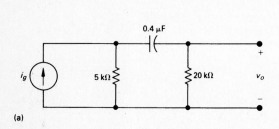

(a)

FIGURE P114

115. [14.68]

   a) Find the impulse response of the circuit in
Fig. P115(a) if v_g is the input signal and v_o is the
output signal.

   b) Assume that the voltage source has the
waveform shown in Fig. P115(b). Use the
convolution integral to find v_o.

   c) Plot v_o for $0 \leq t \leq 2$ s.

   d) Does v_o have the same waveform as v_g?

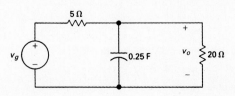

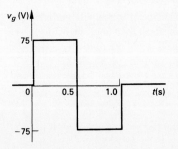

FIGURE P115

116. [14.78] The parallel combination of R_2 and C_2 in the circuit in Fig. P116 represents the input circuit to a cathode-ray oscilloscope (CRO). The parallel combination of R_1 and C_1 is a circuit model of a compensating lead that is used to connect the CRO to the source. There is no energy stored in C_1 or C_2 at the time that the 10-V source is connected to the CRO via the compensating lead. The circuit values are $C_1 = 5$ pF, $C_2 = 20$ pF, $R_1 = 1$ MΩ, and $R_2 = 4$ MΩ.

a) Find and plot v_o to verify your result.

b) Find and plot i_o to verify your result.

c) Repeat parts (a) and (b), given that C_1 is changed to 80 pF.

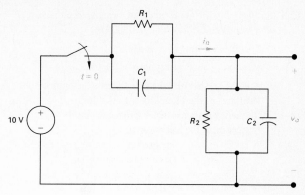

FIGURE P116

117. [15.2] Use a 5 mH inductor to design a low-pass *RL* passive filter with a cutoff frequency of 1000 Hz.

a) Specify the value of the resistor.

b) Plot the frequency response.

c) A load having a resistance of 270 Ω is connected across the output terminals of the filter. Plot the resulting frequency response and identify the cutoff frequency of this loaded filter.

118. [15.3] A resistor, denoted R_1, is added in series with the inductor in the circuit in Fig. 15.4(a). The new low pass filter circuit is shown in Fig. P118.

a) Find the expression for $H(s)$, where $H(s) = V_o/V_i$.

b) Plot the frequency response of $H(s)$ and identify the frequency at which the value of $H(j\omega)$ is maximum.

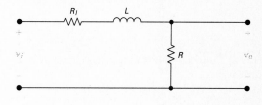

FIGURE P118

119. [15.6] Use a 0.5 μF capacitor to design a low-pass passive filter with a cutoff frequency of 50 krad/s.

a) Specify the value of the filter resistor.

b) Plot the frequency response of the filter.

c) Assume that the cutoff frequency cannot increase by more than 5%. Find the smallest value of load resistance that can be connected across the output terminals of the filter.

d) If the resistor found in part (c) is connected across the output terminals, plot the frequency response of the resulting filter.

e) From the plot in part (d), what is the magnitude of $H(j\omega)$ when $\omega = 0$?

120. [15.9] Using a 100 nF capacitor, design a high-pass filter with a cutoff frequency of 300 Hz.

a) Specify the value of R in kΩ.

b) Plot the frequency response of the filter.

c) A 47 kΩ resistor is connected across the output terminals of the filter. Plot the resulting frequency response and identify the cutoff frequency of the loaded filter.

121. [15.14] Use a 5 nF capacitor to design a series RLC bandpass filter, as shown in Fig. 15.19(a). The center frequency of the filter is 8 kHz and the quality factor is 2.

a) Specify the values of R and L.

b) Plot the frequency response of the filter and identify the cutoff frequencies and the bandwidth of the filter.

122. [15.22] The parameters in the circuit in Fig. P122 are $R = 2400\,\Omega$, $C = 50$ pF, and $L = 2\,\mu$H. The Q of the circuit is not to drop below 7.5. What is the smallest permissible value of the load resistor R_L?

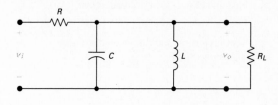

FIGURE P122

123. [15.25] Use a 0.5 μF capacitor to design a bandreject filter, as shown in Fig. P123. The filter has a center frequency of 4 kHz and a Q of 5.

a) Specify the numerical values of R and L.

b) Plot the frequency response and identify the cutoff frequencies and the bandwidth.

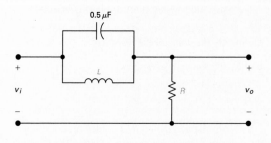

FIGURE P123

124. [15.34] Consider the following voltage transfer function:

$$H(s) = \frac{V_o}{V_i} = \frac{10^8}{s^2 + 3000s + 10^8}.$$

a) At what frequencies (in rad/s) is the ratio of V_o/V_i equal to unity?

b) At what frequency is the ratio maximum?

c) What is the maximum value of the ratio?

125. [15.36]

a) Find the resistance looking into terminals a,b of the circuit in Fig. P125.

b) Find the power loss through the network, in decibels, when the output power is the power delivered to the 50 Ω resistor.

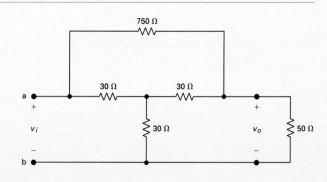

FIGURE P125

126. [16.4]

   a) Using the circuit in Fig. 16.1, design a low-pass filter with a passband gain of 10 dB and a cutoff frequency of 1 kHz. Assume that a 750 nF capacitor is available.

   b) Plot the frequency response of the resulting filter.

127. [16.5]

   a) Use the circuit in Fig. 16.4 to design a high-pass filter with a cutoff frequency of 8 kHz and a passband gain of 14 dB. Use a 3.9 nF capacitor in the design.

   b) Plot the frequency response of your filter.

128. [16.22]

   a) Using 0.1 μF capacitors, design a first-order broadband active filter that has a lower cutoff frequency of 1000 Hz, an upper cutoff frequency of 5000 Hz, and a passband gain of 0 dB. Use prototype versions of the low-pass and high-pass filters in your design.

   b) Plot the frequency response of your filter.

129. [16.30]

   a) Using 10 nF capacitors and ideal op amps, design a high-pass unity-gain Butterworth filter to meet the following specifications: a cutoff frequency of 2 kHz and a gain of at least −48 dB at 500 Hz.

   b) Plot the frequency response of the filter you designed.

130. [16.33]

   a) Using 1 kΩ resistors and ideal op amps, design a low-pass unity-gain Butterworth filter that has a cutoff frequency of 8 kHz and is down at least 48 dB at 32 kHz.

   b) Plot the frequency response of this filter.

131. [16.34] The high-pass filter designed in Problem 129 is cascaded with the low-pass filter designed in Problem 130.

   a) Plot the frequency response of the resulting filter.

   b) Identify the cutoff frequencies, the mid-frequency, and the Q of the filter.

132. [16.37]

   a) Design a broadband Butterworth bandpass filter with a lower cutoff frequency of 500 Hz and an upper cutoff frequency of 4500 Hz. The passband gain of the filter is 20 dB. The gain

should be down at least 20 dB at 200 Hz and 11.25 kHz. Use 15 nF capacitors in the high-pass circuit and 10 kΩ resistors in the low-pass circuit.

   b) Plot the frequency response of the resulting filter.

133. [17.9] Find the Fourier series of the periodic functions in Fig. P133.

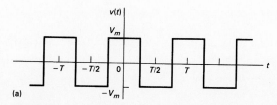

(a)

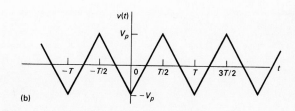

(b)

FIGURE P133

134. [17.22] The square-wave voltage in Fig. P134(a) is
applied to the circuit in Fig. P134(b).

a) Find the Fourier series representation of the
steady-state current i.

b) Use the first 10 terms of the Fourier series
representation of i to plot i, and compare this
with a plot of i determined by analyzing the
circuit.

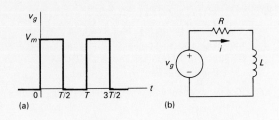

FIGURE P134

135. [17.27] The full-wave rectified sine wave voltage in
Fig. P135(a) is applied to the circuit in Fig. P135(b).

a) Find the first three terms in the Fourier series
representation of i_o.

b) Using plots, how many terms are needed in the
Fourier series to get a good approximation for
the steady-state waveform of i_o?

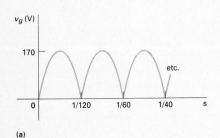

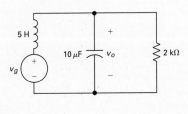

FIGURE P135

136. [17.43] The periodic voltage source in the circuit in
Fig. P136(a) has the waveform in Fig. P136(b).

a) Find the values for C_k, $k = 0, \pm1, \pm2, \pm3$, and
±4 for v_g if $v_m = 54$ V and $T = 10\pi$ μs.

b) Use the complex coefficients from part (a) to
estimate the average power delivered to the
250 Ω resistor.

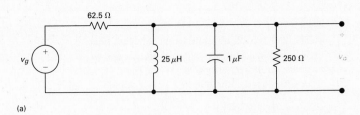

(a)

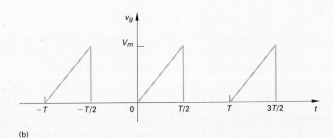

(b)

FIGURE P136

137. [18.34] The input current signal in the circuit in Fig. P137 is

$$i_g = 30e^{-2t} \ \mu A; \ t \geq 0^+.$$

What percentage of the total 1-Ω energy content in the output current signal lies in the frequency range 0 to 4 rad/s?

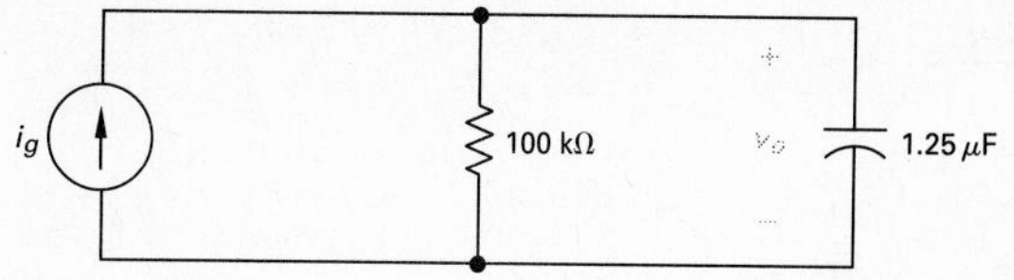

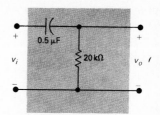

FIGURE P137

138. [18.37] The amplitude spectrum of the input voltage to the high-pass RC filter in Fig. P138 is

$$V_i(\omega) = \frac{200}{|\omega|}, \quad 100 \leq |\omega| \leq 200 \text{ rad/s};$$

$$V_i(\omega) = 0 \quad \text{elsewhere.}$$

a) Plot $|V_i(\omega)|^2$ for $-300 \leq \omega \leq 300$ rad/s.

b) Plot $|V_o(\omega)|^2$ for $-300 \leq \omega \leq 300$ rad/s.

c) Calculate the 1 Ω energy in the signal at the input of the filter.

d) Calculate the 1 Ω energy in the signal at the output of the filter.

FIGURE P138

139. [18.38] The input voltage to the high-pass RC filter circuit in Fig. P139 is

$$v_i(t) = Ae^{-at}u(t).$$

Let α denote the cut-off frequency of the filter, that is, $\alpha = 1/RC$.

a) What percentage of the energy in the signal at the output of the filter is associated with the frequency band $0 \leq \omega \leq \alpha$ if $\alpha = a$?

b) Repeat part (a), given that $\alpha = \sqrt{3}a$.

c) Repeat part (a), given that $\alpha = a/\sqrt{3}$.

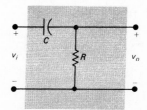

FIGURE P139

140. [19.4] Find the *g* parameters for the circuit in Fig. P140.

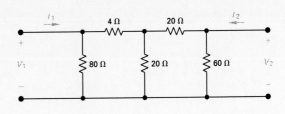

FIGURE P140

141. [19.7] Find the *y* parameters for the circuit in Fig. P141.

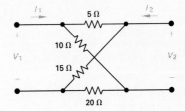

FIGURE P141

142. [19.11] The table below contains direct-current measurements that were made on the two-port network in Fig. P142. Calculate the *h* parameters for the network.

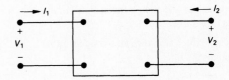

FIGURE P142

TABLE FOR 142

PORT 2 OPEN	PORT 2 SHORT-CIRCUITED
$V_1 = 8$ mV	$V_1 = 5$ V
$I_1 = 0.4\ \mu$A	$I_1 = 5$ mA
$V_2 = -8$ V	$I_2 = 250$ mA

143. [19.19] The op amp in the circuit in Fig. P143 is ideal. Find the *h* parameters of the circuit.

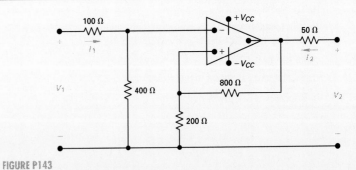

FIGURE P143

144. [19.30] The *a* parameters of the amplifier in the circuit in Fig. P144 are $a_{11} = a_{22} = 0.50$, $a_{12} = j10 \, \Omega$ and $a_{21} = j75 \, m\mho$. Find the ratio of the output power to that supplied by the ideal voltage source.

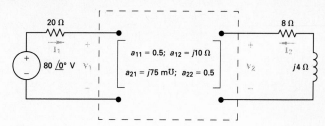

FIGURE P144

145. [19.37] The table contains dc measurements that were made on the resistive network shown in Fig. P145. A variable resistor R_o is connected across port 2 and adjusted for maximum power transfer to R_o. Find the maximum power.

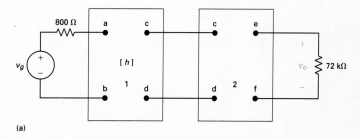

FIGURE P145

TABLE FOR 145

MEASUREMENT 1	MEASUREMENT 2
$V_1 = 4 \, V$	$V_1 = 20 \, mV$
$I_1 = 5 \, mA$	$I_1 = 20 \, \mu A$
$V_2 = 0 \, V$	$V_2 = 40 \, V$
$I_2 = -200 \, mA$	$I_2 = 0 \, A$

146. [19.39] The *h* parameters of the first two-port circuit in Fig. P146(a) are:

$$h_{11} = 1000 \, \Omega; \quad h_{12} = 5 \times 10^{-4};$$

$$h_{21} = 40; \text{ and } h_{22} = 25 \, \mu V.$$

The circuit in the second two-port circuit is shown in Fig. P146(b), where $R = 72 \, k\Omega$. Find V_o if $V_g = 9 \, mV(dc)$.

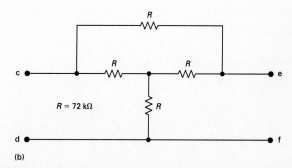

(a)

(b)

FIGURE P146